P9-CDL-942

# PHYSICS FOR ROCK STARS

# PHYSICS for ROCK STARS

## MAKING the LAWS of the UNIVERSE WORK FOR YOU

CHRISTINE McKINLEY

A Perigee Book

**A PERIGEE BOOK**
**Published by the Penguin Group**
**Penguin Group (USA) LLC**
**375 Hudson Street, New York, New York 10014**

USA • Canada • UK • Ireland • Australia • New Zealand • India • South Africa • China

penguin.com

A Penguin Random House Company

PHYSICS FOR ROCK STARS

Copyright © 2014 by Christine McKinley
Penguin supports copyright. Copyright fuels creativity, encourages diverse voices,
promotes free speech, and creates a vibrant culture. Thank you for buying an authorized
edition of this book and for complying with copyright laws by not reproducing, scanning,
or distributing any part of it in any form without permission. You are supporting writers
and allowing Penguin to continue to publish books for every reader.

PERIGEE is a registered trademark of Penguin Group (USA) LLC.
The "P" design is a trademark belonging to Penguin Group (USA) LLC.

ISBN: 978-0-399-16586-3

An application to catalog this book has been submitted to the Library of Congress.

First edition: June 2014

PRINTED IN THE UNITED STATES OF AMERICA

12   11   10   9   8   7   6

*Text design by Tiffany Estreicher*
*Illustrations by Mark Nerys*

While the author has made every effort to provide accurate telephone numbers,
Internet addresses, and other contact information at the time of publication, neither
the publisher nor the author assumes any responsibility for errors, or for changes that
occur after publication. Further, the publisher does not have any control over and does
not assume any responsibility for author or third-party websites or their content.

Most Perigee books are available at special quantity discounts for bulk purchases for sales
promotions, premiums, fund-raising, or educational use. Special books, or book excerpts, can also
be created to fit specific needs. For details, write: Special.Markets@us.penguingroup.com.

For Chuck

## AUTHOR'S NOTE

In passages where it was not possible to get permission to use the names of friends and teachers, I have changed them to protect their identities. Otherwise, the biographical details are accurate.

I have included scenarios and experiments that are obviously dangerous because I know you, the reader, are smart enough not to actually try them. Thank you for considering your safety and using these as thought experiments only.

# CONTENTS

# INTRODUCTION

Physics is the sexiest of the sciences. Sure, you could argue that biology is all about reproduction and chemistry has an intrinsically hot name, but when you get down the guiding principles of the universe, it's all physics. The laws of motion, energy, gravity, and entropy rule. Literally. They trump all other laws and inform all other activity. That's what makes physics sexy. It is firmly in charge.

It pays to be on the physics team. When you apply the basic laws of motion, energy, and gravity to your life, you will become smarter, more successful, and better looking than you already are. I know. It's hard to imagine how you will handle that with graceful humility, because you are already so darn adorable and clever, but physics can help you do that, too. A solid understanding of the laws of physics will not only help you execute a flawless stage dive and win a fist fight on top of a train, but it will also help you cultivate a balanced and sane personal life. I promise.

If you aren't feeling particularly smart, successful, or sexy at this point in your life, the dependable laws of physics will be a great help to you. They offer structure, a safe place to land, and proof that this business of life can be figured out and navigated. That is the beginning of true comfort, grace, and success. My faith in physics comes from experience. The laws of physics guided me from a nervous seventh grader who smoked nonfiltered Camels and swore like an oil pipeline welder, to a career as an engineer, writer, TV host, and musician.

I totally understand if you didn't learn physics in high school. It is a very busy time in which you were experiencing changes in your body and learning many life skills, like how to calculate compound interest, make a left turn at an intersection, and French kiss—the latter two at the same time. It's hard to find the mental space to pay attention in physics class. Even if you were paying attention, the teacher was certainly not talking about anything that related to your goals. You wanted to be a drummer, secret agent, or runway model. You didn't realize that the laws of physics really could prepare you for any career, even the ones without a brochure in the guidance counselor's office.

## It's Not Too Late for Your Science Crush

Appealing role models in science, math, and engineering can be a strong motivator for young people to choose scientific careers. I was lucky to have science teachers who lectured with the passion of traveling evangelical preachers. If you missed out on the chance to bond with a physics-loving mentor, it's not too late. There are plenty of great scientists to worship and imitate.

First and foremost, of course, is Einstein. He traveled the world, challenged Fascists and religious zealots, and was quite the ladies' man. And you may not know them all, but the other stars of physics are just as cool. The fiercely independent Galileo, opium chic Newton, rebellious Marie Curie, hilarious Feynman, and underwear model Niels Bohr.* They didn't become famous, win Nobel Prizes, or have much younger lovers just because they were brilliant and devoted to their work. I believe they were successful because they understood and followed the laws of physics in every part of their lives. They understood that the playbook of the universe is already written out in detail. Studying this playbook was their passion.

So that you may better emulate these glamorous geniuses, I will help you understand and live by the laws of physics. In each chapter, I'll explain a physics concept the way it was introduced to me by my wonderful, quirky teachers at Carondelet High School, a Catholic institution for girls run by the enthusiastic Sisters of St. Joseph. Then I will show you how to apply that concept to your life as a rock star, secret agent, UN sniper, or Roller Derby MVP. I am qualified to do this because I'm a mechanical engineer. That's what we do. We take scientific concepts and make them useful.

---

* There is no proof that Niels Bohr was an underwear model. He did, however, have lovely posture and a full, pouty mouth.

## Scientists and Engineers: Angels and Mercenaries

In the days of Da Vinci, there wasn't a distinction between a scientist and an engineer. (Heck, in Da Vinci's day there wasn't even a distinction between scientist, artist, philosopher, and discus thrower.) These days, our educational and economical systems force us to pick a track—scientist or engineer. We can always jump the track when we feel like it, or take side trips into one another's fields, but in general there is a division between these two disciplines. Here is a shortcut to understanding the difference:* Scientists are tickled to explore the laws of nature without knowing exactly what they're looking for. With their childlike wonder, white lab coats, and prolonged virginity they are a choir of awkward, pale angels. Engineers are the mercenaries of physics. We take the knowledge discovered by these dear scientists and make things humanity really needs—car stereos, panty hose, and guided missiles.

There are mechanical engineers, electrical engineers, structural engineers, and civil engineers, among others. But we all started out as scientists. We had to learn the basic laws of energy, gravity, entropy, and motion first. As I learned these laws, I came to trust them more than I trusted any person or philosophy. I searched the laws of physics for models to inform my personal

---

* Disclaimer: This definition is authored by an engineer and is terribly biased. Scientists might describe themselves as brilliant purists pursuing the absolute truth while engineers are interested only in making money because they are shallow, soulless, and do not truly care about the progress of mankind. As an engineer, I will respond: that is not entirely true. Yes, I am interested in making money, but not exclusively. I also want absolute power.

decisions. I became braver, more confident, and a lot more inter-esting to talk to at parties.*

## Our Physics Mission

Life can be so flimsy and flaky. People can be downright maddening. Gravity, motion, energy, and matter, on the other hand, behave in consistent and measureable ways. Understand-ing them can give you a firm footing in a squishy world, a place to stand while you steer through the dizzying options. Once you know how to use the laws of physics to understand turning gears and spinning planets, you can apply those laws to your personal life. Bonding atoms can provide a useful model for understand-ing romantic attraction. Water boiling into steam can remind you to be patient during life's phase changes. Floating objects can show how to create your own buoyancy. Momentum-conserving crashes suggest the best way to stay on track. The laws of physics provide a beautiful, organized framework for decisions and a comforting sense that this busi-ness of life is not entirely a game of chance.

> There is no better model for our lives than the laws of physics.

There is no better model for our lives than the laws of physics. The atoms we are made of must, quite literally, follow them. If you try to follow a different set of rules, you will be scattered and

---

* If you like talking about the laws of thermodynamics.

frustrated. You can't break the laws of physics, but they can break you. But the laws of physics won't fight you if you don't fight them. In fact, if you understand and work with them, the laws of physics will hold their little lighters up and sing along while you play the rock anthem that is your beautiful life.

# PHYSICS FOR ROCK STARS

# 1

# TEST YOUR HYPOTHESIS
## THE SCIENTIFIC METHOD

At Wendler Junior High in Anchorage, Alaska, I conducted my first science experiment. My seventh grade science teacher, Mr. Daniels, explained the scientific method by writing on the board:

1. Ask a question.
2. Do background research.
3. Construct a hypothesis.
4. Test with an experiment.
5. Analyze the results and draw a conclusion.
6. Report the results.

Mr. Daniels then drew a dashed line with an arrow pointing from step number five to step number three. If your conclusion didn't match your hypothesis, he explained, squinting through his Buddy Holly glasses, you needed to go back and construct a new hypothesis. He showed us a picture of Gregor Mendel and told us the story of the gardening monk who patiently mated peas and tracked the traits of their pea offspring. We looked at pictures of purple and white flowering peas. They had wrinkled or smooth pods with green or yellow peas inside. We drew diagrams of the many offspring options, given their recessive and dominant traits. After peas, we graduated to the offspring options from different combinations of albino mice and spotted cows.

The animal life I wanted most to understand, though, were the seventh graders of Wendler Junior High. I'd recently become aware of a caste system. Cool kids were on the very top and smart kids were near the bottom. I wasn't sure in which genus I belonged. It was understood that you could not be both. In this particular science experiment, this constraint is what we call a "known."

The question of my coolness vs. smartness wasn't simple because, on the one hand, I was in the special advanced math and English classes, played clarinet, and got a special recognition award in the city science fair for my experiment showing the effect of cigarette smoke on house plants. On the other hand, I had some cool-kid cred. I wore awesome jeans with yellow stitching on the back pockets and I knew how to use the F-word as a noun, verb, and adjective.

Despite my ability up to that point to have a firm footing in both camps, it was clear to me that I had to choose: be cool or be smart. High school was soon approaching; the decision had to be made. So, like a good junior scientist, I completed the first steps of the scientific method.

1. Ask a question: Is it better to be a cool kid or a smart kid?
2. Do background research: the cool kids enjoyed upper-caste privileges such as prime lunchroom seating, sexy slow dancing at the after-school functions, and bulletproof confidence.

That's what I really wanted. I wanted to be confident.

I was scared of so many things. My mom started having mysterious seizures when I was in fourth grade. By then my dad had left the house and moved to Texas. I was terrified that a seizure would strike her while she was in the bathtub or driving, that I'd get a call to come to the principal's office and the secretary would tell me my mom had died. I was also scared that her new boyfriend, Chuck, whom I was starting to like, might decide we were too much for him. Worse, Mom could run away with him and leave me behind like my dad had done.

Being cool and unafraid sounded like a relief. I couldn't control any of that other stuff, but I could do my best to not care about it. Whatever was happening at the homes of the cool kids, they didn't seem bothered by it. The coolest girl in school, Sarah, laughed with her mouth wide open, her pink Bubble Yum jammed between her tongue and her teeth. She balled up her report card and threw it in the trash with the expert aim of an NBA free thrower. She sauntered in late to class and shrugged when the teacher asked for a reason. To save energy for more important things such as shoplifting, she trimmed words to one syllable, as in "What's your prob?" She seemed very relaxed. It looked like a good way to navigate life, or at least junior high.

3. Construct a hypothesis: it was better to be a cool kid.
4. Test with an experiment: be a cool kid to experience the difference between smartness and coolness.

I knew two things about the genuinely cool kids in junior high: they got bad grades and they smoked. Those were the first performance criteria I needed to fulfill. I dropped out of the gifted math and English program and sat in the back of my new classes with the Sarahs of the school. I cut my words down to single syllables, using "fas" for facetious and "proc" for procrastinate. Nobody knew what I was saying. Still, getting bad grades in math and English was easy enough.

Social studies class proved to be a challenge. We watched black-and-white films about the exports of each Canadian province and made Tlingit totem poles with construction paper. It was not easy to perform poorly at those activities; the D on my midsemester progress report was hard earned. I had only managed it by not turning in my diorama of an Alaskan gold mine town.

Learning to smoke was no piece of cake, either. It wasn't just the smoking that made one cool; it was *hiding* your smoking. I learned, from one of the cooler eighth grade boys, to cup a burning cigarette in my hand and put that hand in my jacket pocket. All this smoking and hiding of smoking was done while skipping gym class. This was effective multitasking since it addressed both performance criteria: smoking and bad grades. At home in front of the mirror, I practiced raising an unlit cigarette casually to my mouth with two fingers. But no amount of practice could keep me from feeling dizzy after inhaling the sickening menthol-tinged smoke of a Winston. It was awful, but I toughed it out. Mendel didn't quit after his first crop of peas turned out to be all one color. I was no quitter, either.

By the time my mom saw my mid-semester report card, Chuck noticed I was stealing his cigarettes. He hid throat-searing Turkish cigarettes in his pack and my mom instituted "home-work time." I was forced to build a sad, miniature mining town from cardboard and twigs to bring up my social studies grade. Bits of glitter gold dust stuck to the muddy streets, and the tooth-pick prostitutes waved wearily from the windows of the tilting saloon. I sympathized with them. I was tired, too, and I was coughing like a tube pashe. (That's cool kid for tuberculosis patient.)

But I soldiered on. I wanted to be a cool kid very badly. I wanted to have some control of my life. Not caring how it turned out was the best strategy available to me.

Then everything changed. My life flipped completely over. Chuck married my mom, adopted my sister and me, and we moved to California. We drove there in our dirty Chevy Blazer, our pasty white faces at the windows, squinting at the sunlight. When we got there, my parents launched a strong counteroffen-sive to cool by enrolling me in Catholic high school.

In my high-waisted blue plaid skirt with big, stiff pleats and white knee socks, there was no way to look anything close to a cool kid. Also, skipping gym class was impossible. The Sisters of St. Joseph were a cohesive stealth unit with the instincts of well-trained marine reconnaissance troops. Their flat, rubber-soled shoes were ideal for quick, silent approaches, and their hearing was supernatural. After sitting in Sister Rosemary's office and listening to her explain to me why swearing in the parking lot is not acceptable because it erodes the learning environment for the whole community of educators and students to which I belonged, I decided not to waste my time resisting the sisters. I could tell they had broken tougher girls than I. Also, I kind of liked them. They had true confidence. They knew how the world worked, and it was their mission to teach me everything they could about it in the four years I was their captive guest.

5. Analyze the results and draw a conclusion.

I could continue trying to be cool, but I didn't make sense anymore. I didn't want to smoke or even skip gym class. I loved running laps on the big lawn in the boxy red regulation shorts, the sun warming my skinny white legs. I knew I didn't look cool, but I didn't care. While the Mother Mary statue looked on with quiet approval, I sprinted out ahead of the other girls, sucking in the smell of wet grass, running as fast as I could from the long Alaskan winters. By then I'd been a cool kid long enough to know their secrets. They weren't any less scared or any more relaxed. They hated going home to fighting parents, drinking mothers, and empty kitchens. That's why they stayed huddled in a group, smoking their cigarettes in the Anchorage snow, watching the last glow die before they walked home to brave the darkness.

In my new school, sun streamed in the windows and rainbow banners ordered me to rejoice. I was in the safe brick and ivy palm of the Sisters of St. Joseph. At home I was no longer in charge of grabbing the steering wheel or pulling Mom out of the tub if she had a seizure. That was Chuck's job. My experiment had blown up. All the constraints had changed. This happened to even the best scientists.

I remembered Mr. Daniels describing Gregor Mendel's results. The robed friar patiently pollinated his peas in the abbey garden, mating white- and purple-flowered parents only to get all purple-flowered progeny. Then, in the next generation, a pure white pea popped up. Not a blended light lavender like Darwin's work suggested, but a pure white flower child from solid purple parents. The gardening friar understood from that one pure white flower that genes were handed down to us from our parents not smashed together, but fully intact. This discovery would help us understand how we inherit our eye color, blood type, and our adorable dimples, but it wasn't what anyone expected. I knew from Mendel and his peas that all this blooming, growing, rethinking, and starting over again were part of the scientific method. I hadn't done anything wrong; my experiment had progressed.

Chuck still left his cigarettes out by the garage stairs, but I didn't steal them anymore. I would come out to the garage when he was smoking and ask him how he lost his finger in Vietnam, how thermometers worked, anything just to hear his low, calm voice and be surrounded by his moat of protective smoke. One day he said, "You're smart. Use it. You can do anything you want." At fourteen years old, all I knew was that I wanted my life to be a glamorous adventure. Chuck and the Sisters of St. Joseph were making it clear that being smart was the way to get there.

It was time to wrap up my experiment.

3. Create a new hypothesis: smart is better; smart is the very best.

# 2

# MAKE SPACE
## NATURE ABHORS A VACUUM

I know that nature abhors a vacuum because Sister Rochelle announced it in the third week of ninth grade. It seemed odd for nature to so desperately dislike anything (or in this case, nothing) so much, but I felt like I could trust Sister Rochelle. The only nuns I had seen before her were in movies in which they were feeding the poor or assisting in exorcisms. In both cases, they showed themselves to be sincere and helpful. Sister Rochelle was a little different from those movie nuns. She wore bright-flowered dresses, had the muscle tone of a small, powerful gymnast, and addressed Jesus out loud before each class in a casual way as though they were exchanging couponing tips over coffee. Still, she was a nun, so I had no reason to doubt her integrity.

Sister Rochelle explained that a vacuum is pure emptiness. She spread her fingers into little sunbursts, shot her arms over her head, and waved them through the air. This jazz dance move was to illustrate the abundance of oxygen, nitrogen, and helium atoms in the air around us. Sister Rochelle explained that we are so accustomed to these invisible air atoms and the pressure they put on our bodies that we don't even notice them anymore. If that air were to disappear, though, along with its 14 or so pounds per square inch of pressure, we'd be in trouble. Our eardrums would rupture, our internal organs would expand, and we would find ourselves uncomfortably bloated and, soon after that, uncomfortably dead.

After she dropped her jazz hands, Sister Rochelle tilted her head, pointing her cowlick at us. This indicated that it was time to reflect on atmospheric pressure. That is one thing I noticed right away about my new school: there was a lot of reflecting. In one single school day, for example, we might be asked to reflect on New Testament parables, the symbolism in *The Old Man and the Sea*, and what kind of message we sent if we did not wear a bra under our white uniform blouse. For the latter, my first guess was not correct. Apparently, going braless does not say "my whites were still in the dryer when I left for school" but instead announces "I am a sexually reckless young woman destined for a life of syphilis-induced blindness."

To nudge us into scientific reflection, Sister Rochelle would ask a question and then raise her eyebrows, look at us eagerly, and lean toward us with breathless expectation. It was the breathless part that worried me. If no one answered, it seemed very possible that she could pass out and tip over on the front row. None of the other girls seemed worried. Maybe they knew that this was common nun behavior. They knew a lot I didn't, crossing themselves on an invisible cue, doing that weird little curtsy before sitting in a pew, standing up, sitting down, and responding "amen" and "also with you" in spooky unison. I was consistently two seconds behind everyone else, crossing myself backward, standing when I should be sitting, mumbling "doe ray me doo" and hoping it would rhyme with whatever they were saying.

For all those girls knew about the dance steps of Catholicism, they lacked an understanding of the body's need for respiration. Or they didn't care. I answered as well as I could while the other girls sat in murderous silence, perfectly content to watch our teacher play a deadly game of blue-faced charades. We worked through the entire vacuum lesson like this—me doing my best to answer the tiny sister's questions and unfinished statements about air molecules to keep her from running out of air herself.

To demonstrate nature's abhorrence of a vacuum, Sister Rochelle narrated over the noise of a small electric pump connected to a plastic water bottle by a tube. As the pump sucked air out of the bottle it shrunk and twisted in agony. Sister Rochelle continued her torture and cheerfully explained that as we removed the air on the inside of the bottle, the air on the outside pushed against the bottle's plastic. With no air on the inside anymore to push back, the bottle felt pressure on the outside. Scientists say that "nature abhors a vacuum" because it seems the air on the outside will do anything to get into that empty space inside the bottle.

To illustrate the air pressure on the outside of the bottle, Sister Rochelle made a sudden windmill karate move with one of her arms, stopping just short of the crumpled plastic. I reflected on how I'd underestimated the sister's street-fighting skills, given her size and vocation and how she might be great backup in bad neighborhoods. Then I reflected on how I was getting the hang of reflecting.

When Sister Rochelle opened a bag of giant campfire marshmallows, she had my full attention. If she lit up a Bunsen burner and whipped out some graham crackers and chocolate bars, I wanted to be the first to volunteer as her assistant. I surveyed the room and found that I was well positioned. No one else was tracking this marshmallow thing.

Sister Rochelle placed three marshmallows under what looked like a big, overturned glass mixing bowl and started the pump's thumping motor again. A cork stopper was stuck in a hole at the top of the bowl. A tube reached through the cork stopper on one side and ran to the pump on the other. When she asked us what we thought would happen when she pumped air out of the bowl, she was met with silence as she breathlessly waited for an answer. As usual, Sister Rochelle and I worked through it together. Never mind the whole "community of young Christian women" thing. I was in this alone.

After establishing that the glass bowl wouldn't crumple or crack because it was stronger than the plastic bottle, I noticed the marshmallows puffing up. With no air around the marshmallows, there was nothing pushing on them. They expanded into the empty space. Pretty cool. Sister Rochelle was on a roll now. She made sure to tell us that if we ever stuck our head in an astronaut's helmet and pumped all the air out of it, our faces would swell up just like those marshmallows. We'd have about 90 seconds to reflect on the magic of science before we were asphyxiated. During that time, we might feel the saliva on our tongue boil from the lowered pressure. She reminded us that screaming would be of no use since sound is a wave that travels by compressing air, so it can't travel in a vacuum. Noted. I will never, ever put my head in an astronaut's helmet and pump all the air out of it to see if my face swells up.

The next week, as Sister Rochelle progressed from experiments with air to experiments with water, I learned a few more important lessons. The first was not to stand too close to her if she is holding a lab water hose because she gestures with her hands quite a bit. The second is that water behaves just like air in many ways. When water surrounds a waterless pocket of the universe, it fights to get inside that dry pocket—just like the air on the outside of those plastic bottles fought to get in the airless space inside them.

Sister Rochelle explained that we all use this principle when we sip through a straw. It may feel like we are pulling ice tea into our mouths, but all we are doing is evacuating some air from the straw so the tea can get a push into it. That push comes from the air on the surface of the tea sitting in your glass. The tea hanging out in your glass feels the weight of the air molecules (atmospheric pressure) hovering over it. Since you have sucked the air out of the straw, the pressure on the tea in the glass is suddenly higher than the tea in your straw, and up goes the green tea ginger spritzer into your mouth.

## Betting on the Limits of Atmospheric Pressure

Here's a fun way to make a little extra spending money. Sister Rochelle didn't tell us to do this, exactly, but she armed us with the knowledge to pull it off. Next time you are backstage with more than one band after a show, tell the lead singers you want to see which of them has the strongest lungs. (Lead singers are incapable of resisting a competition of any kind and they are very proud of their lung capacity.) Propose that they each bet twenty dollars on a lung strength challenge (multiply this by 10 if their band is traveling in a luxury tour bus, multiply by 100 if only the gear goes by bus and the band travels by plane). Tell them the winner gets all the money and if neither succeeds in the challenge you will keep the money. While both lead singers are warming up their lungs with high notes, pull two 30-foot lengths of straw-sized plastic tubing from your purse. (There are other excellent reasons to have plastic tubing on hand at all times.) Next, borrow two ladders from the lighting tech. Have each lead singer climb an equal number of rungs on the ladder—enough that their heads are 30 feet from their beer bottles below. Get out a stopwatch and tell the lead singers that the first one to spit beer from their mouth like a Grecian fountain, showing that they have successfully sucked the beer through their straw, is the winner. Make a big deal of counting off the start, and then watch as they suck their cheeks in like supermodels. When neither of them manages to collect any beer from the bottles below, sweetly wish them better luck next time, tuck your winnings into your pocket, and make a quick exit.

You can execute this scientific demonstration ("con" is such an ugly word) with absolute certainty of the outcome because no matter how strong the singers' lungs are, they can only evacuate the air from the straw. Since there is nothing pulling the beer up, just air pressure pushing from the other side, there's a limit to how high the beer will rise. And the limit just happens to be around 30 feet. You win.

Hand pumps work in a similar way. The pumping evacuates

the air inside the pump and water whooshes in, pushed by the air on the top of the well, lake, or in the perfume bottle. Just like the straw, the hand pump isn't doing any pulling. It is simply clearing out air molecules so that the air weighing down the other side of the liquid can win the shoving match.

## The Physics of Life: Working Nature's Compulsion to Fill a Void

I find it necessary to keep a close eye on nature's abhorrence of a vacuum when my life is overstuffed and I start to clear out space in my closet and on my calendar. I know that nature will immediately try to fill it. She's compulsive that way. And since I know that nature will fill space with anything available, it's important to take nature on headfirst. I take control of both the vacating and the filling. If I vacate Friday nights and fail to fill them, nature will do it for me. She is not picky. She'll use whatever is nearby to fill space. I will be in danger of getting recruited into a beer polo league or talked into providing exotic bird temporary foster care services. Both are perfectly acceptable activities, but I am a major lightweight and birds scare the life out of me. (I saw a few choice scenes from Hitchcock's *Birds* when I was about three years old. Big mistake.)

Same goes with a vacuum in one's dating life, career plans, or any other area of life that requires a bit of thoughtful monitoring. It is up to us to fill the space before nature does. After I've cleaned the spaces out of my life, I must be the bitch that nature is, take a page from her playbook, and fill the space quickly and fiercely with something I can get my hands on.

Now, if you don't feel like filling the new spaces you've created right away and you need time to decide precisely what you want, get ready for some serious pushback from nature. It's possible to keep the space open, but it will take focus. Some monks and yogis spend their entire lives creating space. They may look like quiet, humble people but they are experts at slap-fighting nature. Indeed, they are pushing back a flood of distractions, doubts, and desires while nature tries to crush them like one of Sister Rochelle's plastic bottles. When nature presses in on the space you want to preserve, do what I do: picture Sister Rochelle holding the hose of a vacuum pump near your heart, your head, whatever needs clearing out. Close your eyes and think of nothing but the thump, thump, thump of the vacuum pump.

## ((( PHYSICS PRACTICE )))

1. Atmospheric pressure at sea level is 14.7 pounds per square inch. That pressure can push water up a completely empty tube 34 feet. Why, then, can a person sucking on a straw only get water to rise about 20 feet?

ANSWER: Our mouths aren't capable of creating a perfect vacuum. No matter how much we pucker and inhale, we can't get every atom of air out of the straw. Because of that, there is still some air pressure on the straw side pushing against the atmospheric pressure that is pressing on the liquid's surface. So instead of the liquid rising 34 feet, it rises 20 feet or so. Since humans typically like to keep their drink within arm's reach, this straw-drinking limitation is usually not a problem.

You can calculate this on your own because you remember the density of water from high school science. You were totally paying attention and not at all distracted by the weird shape of the kid's neck in front of you. Water is about 62.4 pounds per cubic foot. Put weight of the water on one side of an equation and weight of the atmospheric pressure (converted to pounds/ft$^2$) pushing down on the other side and solve for the height:

62.4 pounds / ft$^3$ × height of water = 2117 pounds/ft$^2$
Height of water = 2117 / 62.4 feet
Height of water = 34 feet

**2. When marshmallows expand in a vacuum, does their weight change? Do they have more calories?**

ANSWER: The weight and caloric content of the puffier marshmallows stay exactly the same because no marshmallow mass was added or subtracted. The mass only spreads out differently.

**3. If the highest you can suck water in a straw is 18 feet and you try again with a smoothie that has a density that is 1.2 times that of water, will you be able to suck the smoothie higher in the straw, or not as high as the water?**

ANSWER: Since you have the same amount of atmospheric pressure pushing on the surface of the beverage, you won't get any additional help from that side. On the straw side, your mouth will create the same pressure differential between the surface of your drink and the straw. The only thing that changed is the weight of the liquid in the straw. Since the smoothie is heavier than water, it won't rise as high as the water did in your straw.

**4. Who uses the word "abhor" anymore?**

**ANSWER:** Science teachers, vampires, and Civil War reenactors.

## ((( TRY THIS! )))

We know, from the bet with the singers, that drinking from a 30-foot straw is not possible, but let's find out how long a straw one can drink from and how atmospheric pressure contributes to that. Yes, let's!

You'll need: Two 25-foot straws, two mojitos, one team of competitive high school or college cheerleaders.

On a sandy beach (to assure that you are at sea level and to give yourself a soft place to land), place the mojito glass in the sand. Get a group of incredibly strong and supernaturally balanced cheerleaders to lift you into a seated pose at the top of a standing pyramid. Try to drink the mojito. Cut 1 foot from the straw and change formations to lower your seated pose by 1 foot. Repeat until you can drink the mojito. Note the longest straw you could drink through. Then bring the mojito and the cheerleaders to the foothills of the Himalayas. Repeat. What's the longest straw you were able to drink through in the Himalayan foothills? Was it shorter or longer?

**ANSWER:** On the beach, with the full force of atmospheric pressure pushing down on the surface of the mojito, when you suck the air out of the straw, the mojito will rise in the straw 10 to 20 feet, depending on how good you are at sucking the air out of the straw.

With the lowered atmospheric pressure in the high-altitude Himalayas, there will be less pressure to push on the surface of the mojito, so the longest straw you were able to drink through there will be shorter than your beach straw, even if you are just as good at sucking the air out of the straw. There is simply less pressure on the surface of the mojito in the glass to push it up the straw.

Treat the cheerleaders to pink Himalayan salt scrubs. They have been more than patient with your bizarre requests.

# 3

# KEEP IT REAL
## THE NECESSITY OF NUMBERS

In my sophomore year of high school, I dreaded algebra class. The parabolas, natural logs, and imaginary numbers flew over my head in a twisted cloud. If there was a way to get through life without knowing any more arithmetic, I wanted to find it.

I didn't yet see the elegance and usefulness of numbers. Not until years later did I realize that math was the only language we could use to accurately describe bacteria growth, air pressure, and waterfalls. With math we make the leap from appreciating conservation of energy as a concept to using it to keep a bridge from collapsing. Math moves us from living in a cave that we happen to find to designing a home that will safely perch on a hill. We become inventors rather than scavengers, designers rather than slaves to trial and error.

It's possible to talk about physics and leave out math entirely, but it's not easy. It would be much like describing a birthday party to someone not familiar with one without using any nouns. It can be done with a combination of verbs, pronouns, and pantomime, but there is likely to be some lack of clarity about what exactly was set on fire and why the guest of honor was forced to put out the blaze using only his breath.

In the same way that nouns are crucial when describing parties, mathematical expressions are necessary when describing physics. The cleanest, sleekest way to describe floating yachts, speeding bullets, and the angular momentum required for a perfectly executed stage dive is mathematical. For those of you who feel that algebra sucked as much fun from your high school years as your curfew did, I understand your reluctance to revisit math.

I assure you, I'm not naturally a math person, either. I have become capable with numbers, but I didn't start out that way. When I was in first grade, I was very excited about getting older than my sister. I was six, she was nine, so in just . . . oh, I couldn't calculate how many birthdays . . . I would be older than she. I would be her big sister! I wish that I could say I had an early intuitive grasp of the flexibility of the space-time continuum, but that was not the case.

I was a good reader, but not gifted with numbers. In our years of playing Monopoly, my sister easily tricked me out of thousands of dollars of multicolored cash and plastic real estate. I picked a different playing piece each time, thinking that the dog would be more successful than the shoe, the thimble luckier than the hat. Nothing helped. If you can't make change properly for a $100 bill, nothing is going to help you survive in the high stakes world of fake finance—even if you are the sexy sports car. You will still face the embarrassment of being repossessed by an empire represented by a goddamn wheelbarrow as the game grinds to a painful end.

By seventh grade, I was a decent little number-wrangler. But sophomore year, when algebra class got rough, I announced at the dinner table that I was going to drop it. Chuck showed his newly sharpened parental skills by saying that I should talk to the guidance counselor. He knew that she was not going to help me get out of anything. He was right. My and the guidance counselor's conversation was very short. The only thing I got out of it was a note from her excusing my tardiness as I shuffled right back into algebra class. I soldiered on through algebra and managed to pull my C up to a B minus.

In my junior year, my math outlook changed. The stylish Mrs. Johnson taught trigonometry and pre-calculus. With her knee-high boots and long, straight hair, she was a sleek and glamorous mathematics ambassador. She convinced me that triangles and curves were elegant and powerful. When I went to her office for help one day, she guided me through a graphing exercise, and then we chatted about shampoo frequency and she tipped me on the ability of hair to stay glossy when it was washed every other day. The combination of trigonometry and beauty advice made a huge impression on me. Math and beauty were forever entwined in the forming adolescent brain.

By college, calculus made perfect sense. I realized by then that math classes didn't get harder as you progressed. It was like taking a language. The third year of Italian isn't harder than the first year. Italian is easier when you already know how to roll those

$r$'s, or in the case of calculus, picture a line inching nearer and nearer to an axis without ever touching it.

I'm grateful that I plowed ahead with math. I had the mistaken impression that if math seemed difficult, then I just wasn't cut out for it. Now I realize how ridiculous that is. We would never say, "Reading didn't come naturally to me so I stopped doing it." We consider reading to be a crucial survival tool. Why don't we feel the same way about mathematical literacy? To understand our own finances, our medical records, and when someone is offering us a terrible deal on insurance, we need to step through the numbers with clarity and confidence.

## Math's Commanding Grip on Reality

For the purposes of understanding basic physics, let's agree that a mathematical statement is simply a snapshot of reality. For example, $3 + 2 = 5$. Yep. That's true. You could communicate the same thing in words: three plus two equals five. As ideas get more complex, though, using words gets tougher. An equation actually becomes the more intuitive and aesthetic choice. Saying, "The increase or decrease of energy will equal the increase or decrease of mass multiplied by the speed of light squared, which is a constant" is just not as succinct or sexy as $E = mc^2$.

The real fun begins when the mathematical expression turns into a question like this: $3 + x = 5$. The missing number, $x$, has to be 2, right? This kind of numerical question is very useful when you know some basic truths of the universe and need to find the occasional missing piece.

Here is the only rule you need to follow: you must keep it honest. If a mathematical statement starts out true and you mess with one side of the equal sign without doing the same thing to the other side, you will make it untrue. You must do the same thing to each side. If you add 47 to one side, you must add 47 to the other side to keep that equals sign from being a lie. If numbers make you nervous, right now you may be thinking, "Why would I want to mess with either side of the equals sign? Why not back away slowly while avoiding eye contact with the mysterious x?" Well, you might want to get the one thing you don't know (speed, weight, time) all alone on one side of the equation so that the little snapshot of reality described by the equation simplifies to a useful statement such as, "How fast do I need to be going in my car to jump that half-open drawbridge?"

Working on a mathematical statement is much like packing your belongings in a small plane for a quick ride to a fishing lodge in Alaska (where you will be served salmon prepared twelve different ways in your weeklong stay). The pilot needs the weight on each side of the plane to be equal. If, after he announces that the plane is loaded equally on both sides, you realize you have two equally weighted packages of survival essentials yet to be loaded on the plane, do you put these cases of slow-roasted organic coffee beans and hair removal cream for sensitive skin (a whole case, really?) in the overhead compartment on one side of the plane? Of course not; you put one on each side to keep the plane balanced. It was balanced before, and you are keeping it that way by adding the same amount of weight to each side. The overall weight in the plane has increased, but it has increased by the same amount on each side, so the pilot is still happy. And when flying into the backwoods of Alaska, you really do want that pilot to be happy.

## Converting Ideas to Numbers

To measure and tweak the things we're interested in, we need them to be expressed in the form of numbers. We do this all the time when we ask a friend, "On a scale of 1 to 10, how awkward was your date?" The answer to that question is more informative than "Was your date awkward?"

If you were studying a group of college students and wanted to know what they found attractive, you could show them pictures and ask them to tell you how attractive they find the people in the pictures. They would give you a lot of answers, such as "Super hot," "Not so much," "I'm crazy about guys with dragon neck tattoos," and "I like women who look exactly like my mom." As disturbing and entertaining as that information might be, it is not very useful data. You could quantify the responses by asking the students to rate the photos from 1 to 10. You would then get some numbers that would be a little more useful. For an even more precise (and honest) measurement, you could record the physiological responses that indicate attraction. If flared nostrils, rapid eye blinks, and underarm temperature were known indicators of attraction, you could measure those. So, with a nostril flare of 2 millimeters, a 5-per-minute increase in blinks, and an underarm temperature of 98.9 degrees F, you could know exactly how the viewer felt about the photo.

In scientific circles, this process of translating information into numbers is called "quantifying data."

## Units: The Fine Line Between Numbers and Nonsense

Once you have numerical data, you can combine and chop it up in sensible ways. In freshman science class, Sister Rochelle taught us how to keep our units straight. This is crucial. If you start mixing gallons with cubic feet, moles with milligrams, or horsepower with Newtons, you will find yourself in a dangerous wonderland of your own making, and you will soon fall down a rabbit hole of freaky proportions and nonsensical results.

To keep your units in good order, treat them just like numbers they describe. Remember multiplying fractions in school? You can cancel something on the top with something on the bottom, right? So ¼ multiplied by 4 can be rewritten as ¼ times 4/1. If the top is multiplied with the top and the bottom multiplied with the bottom, you get 4/4. The 4 on top cancels the 4 on the bottom and you get a big, fat number 1. (I apologize if this is giving you traumatic flashbacks to sixth-grade math class. Stay with me. Take deep breaths. This time you won't be graded.)

We can cancel units on the top and bottom of fractions in the same way. So if you can run a mile in 8 minutes and you want to know how many minutes it will take you to run a marathon if you keep up that pace, do some unit-canceling:

A mile run in 8 minutes can be expressed as 8 minutes/mile. A marathon is 26.2 miles. So your equation looks like this: 26.2 miles × 8 minutes/mile. Remember, the miles after 26.2 is on top of an invisible number 1. Miles cancel because 1 is on top and 1 is on the bottom; 26.2 ~~miles~~/1 × 8 minutes/~~mile~~ = 209.6 minutes. That's about 3 hours and 30 minutes. Not bad.

That is good, solid, dimensional analysis. You look at the units attached to the numbers and arrange them mathematically so that they cancel to give you the information you are looking for. It's very handy because knowing what units are on top or bottom of the fractions makes it clear how you need to set up the numbers attached to them.

All that said, I promise not to force you to do any algebra in this book. Just know that when I use acceleration to find speed or calculate gallons of water using its density and weight that I'm engaging in some honest math and keeping the laws of the universe intact by following the rules of balancing each side of the equation and watching my units. I'll occasionally show an equation. If these math mileposts don't help you visualize the concept, just treat them like you would a sweaty couple making out at an office holiday party. Take a curious peek when passing, but keep moving, lest they ask you to join in . . . again. I understand if you can't get as passionate about algebra and geometry as the followers of Pythagoras who, rumor has it, killed a member of their cult for using irrational numbers.* They believed, with all their nerdy numerical hearts, that whole integers are the creations of God, and that numbers that can only be expressed with long, unwieldy decimals instead of fractions of whole integers are bad news. I admit they might have been taking an interest in math too far, but I do understand why they were so moved by mathematics.

---

* Irrational numbers are unfairly named. They are not numbers who clearly state "no gifts please" on their birthday party invitation, and then end up all angry with their friends because they didn't bring gifts. They are just numbers such as 1.41421 that can't be neatly described by a fraction of whole numbers. Since 1.5 can be described as 3/2, it's considered rational and nobody in Pythagoras's wackadoodle sect will get hurt for talking about it. Who is really the irrational one here?

The power of trigonometry and geometry to calculate the height of mountains without climbing them and the distance of diagonal shortcuts before taking them must have seemed like a magic trick. They were decoding the secret language of the universe.

The Russian mathematician Sofia Kovalevskaya, who hung out with Charles Darwin and Aldous Huxley at George Eliot's Sunday salons in London, said "It is impossible to be a mathematician without being a poet in soul." She was absolutely right. I would add that it is impossible to be a sucker if you are a mathematician. Even an amateur mathematician can confidently navigate retirement funding, graph weight loss, and compare frequent-flier deals. And when you are done using math for those daily tasks, you can also use it to design a speedboat and analyze the effect of new treatments on cancer cells. At its best, math truly is like poetry—succinct, understated, and powerful.

## ((( MATH PRACTICE )))

1. Match the physics concepts below with the equation. If you've forgotten most of what you learned between sixth and tenth grades, here's a quick refresher: numbers are multiplied if they are right next to each other, in parentheses next to each other, or have an asterisk between them; so these expressions all mean x multiplied by y:

xy

x(y)

x*y

Numbers are divided if they are separated by a slash, so x divided by y looks like this:

x/y

A few other things you'll need to know if you have forgotten everything before tenth grade: Any number divided by 0 is infinity. The length of the circumference (edge) of a circle is 2 × pi × the radius of that circle. Also, the American Revolutionary War was fought against the British. Except for books about brave teenagers in the rugged West, old men in boats, and some gruesome pictures of STDs—that pretty much covers those missing years. Now let's start matching!

## WORD STATEMENTS

1. An object accelerates when force is exerted on it. The force is the mass in motion multiplied by its acceleration.

2. If a body gives off energy in the form of radiation, its mass diminishes by the amount of energy divided by the speed of light squared.

3. The potential energy of a mass at an initial height equals the kinetic energy it will have after falling from that height and converting height to speed using the acceleration due to gravity.

4. There is nothing more annoying than people who start eating a sandwich at a red light and then don't see the light turn green and then when you tap your horn to let them know the light has been green for a while they give you the finger like you are the rude one when really they are totally the rude one.

5. Distance makes the heart grow fonder.

6. Familiarity breeds contempt.

## MATHEMATICAL STATEMENTS

A. $E = mc^2$

B. $F = ma$

C. $mgh = \frac{1}{2} mv^2$

D. Time delay $_{\text{red to green}}$ = Annoyance
   Time delay red to green + Finger = Annoyance/0

E. Contempt = TK + 1/(Cd), where T is hours spent in the same room, K is an empirically derived constant expressed in contempt accumulated per hour, C is the empirically derived constant expressed in Fondness/miles, and d is distance in miles. Note: where miles become small enough that overall Fondness (Cd) becomes less than 1, the function 1/(Cd) becomes greater than 1 and adds to the overall contempt.

F. Fondness = Cd, where C is an empirically derived constant expressed in Fondness/miles and d is distance in miles.

ANSWER: A2, B1, C3, D4, E6, F5

2. Use a mathematical statement to describe calories you would consume and burn on the following evening: After eating a 500-calorie plate of nachos at happy hour, you dance for 27 minutes using 10 calo-

ries/minute, then walk 1.5 miles to a concert using 100 calories/mile while eating 12 gummy bears, which are 9 calories each. At the show you and your friend equally split a pint of beer that is 205 calories a pint. You then must use your 12 calories/minute karate skills for 4.5 minutes to defend your friend because he has a really big mouth and likes to chat with girls who have angry boyfriends. When the police arrive, you and your friend run at a pace that burns 100 calories/mile for 0.5 mile to a roundabout with a radius of 60 feet where you continue running halfway around it before you catch a cab and make it home.

ANSWER: Calories = 500 − 27(10) − 1.5(100) + 12(9) + 205/2 − 4.5(12) − 100(.5 + (3.14)(60/5280)). After all that excitement, you have consumed 186 more calories than you have burned. Jump rope for about 20 minutes before going to bed and you're all evened up.

# 4

# DON'T SPIN
## CONSERVATION OF ENERGY

The same week Sister Eleanor described Jesus raising Lazarus from the dead, Sister Rochelle announced that energy is neither created nor destroyed, but it can change forms. These two lessons seemed related.

I worked through it in my head as Sister Rochelle wrote the equations for potential and kinetic energy on the board. If, as Jesus and the sisters insisted, there was something left of our soul after death, it must leave our bodies. If the soul wasn't a physical thing, it must be a form of energy. As the last bit of soul energy left Lazarus to float up to heaven, Jesus must have snatched those tiny clouds of energy from the air, cupped them in his hands, and gently blew on them to bring them back to life. I imagined Jesus pushing the energy back into Lazarus's chest with a firm shove and the dead man's eyes popping open in surprise. This was energy changing forms, not created, just warmed, massaged, and moved around. Jesus had clearly read the fine print of this first law of thermodynamics because he had this changing forms exercise down. Lazarus was dead one minute, then sitting up and asking for a glass of water the next. Good job, scientist Jesus.

I wondered if Sister Eleanor and Sister Rochelle had worked their lesson plans out together at the convent next door. I pictured them drinking daiquiris and loudly discussing the conversion of death to life, water to wine, and faith to action as they embedded clues in their class material that would help us tie it all together: Jesus as scientist, Newton as savior, God writing out the commandments on one side of the stone tablets and the laws of thermodynamics on the other. It was far-fetched that the sisters would share this larger vision; I knew that. Still, they were the only two staff on the Spirit Week committee. It looked a little suspicious.

We can never create any more energy.

Jesus and Sister Rochelle may have been comfortable with the idea of the universe having all the energy it will ever have, but for me it's still a startling idea. We can never create any more energy. We can only be an agent of energy change. The sun's energy is stored in sugarcane; we eat the sugar and convert its calories into the energy we require for kissing and skinny-dipping. Like every other living thing in the universe, we are energy-conversion machines. You can convert doughnuts to dancing in your beautiful legs and cookies to calculations in your magnificently firing brain, but you cannot create any new energy. We have all the energy we ever will in our universe. That's the first law of thermodynamics. That's right, it's the *law*.

A lot of the energy conversion we experience is from what

we call potential energy to kinetic energy or the other way around. Potential energy is unmoving energy waiting to be somehow used or fired

> We have all the energy we ever will in our universe.

up. Kinetic energy is the energy of motion. When we throw a ball straight up in the air, we give it a handful of kinetic energy (motion energy). The ball pauses in midair before coming back down. At that moment it pauses, it has no kinetic energy and all potential energy. When the ball hits the ground, it has converted its potential energy (height) back to kinetic energy (speed). It spends that kinetic energy in the collision with the ground— making a nice thump, denting the ball in for a moment, and maybe pushing aside some grass and dirt.

## Potential Energy: Crouching Tiger

But who are nineteenth-century physicists, no matter how cool their beards,* to tell us that we can't create energy? A power plant creates energy, right? Nope. Even when we think we're creating energy, we're really just converting it. In a power plant, we might convert natural gas to heat, which turns water into steam, which pushes a turbine, which spins a generator, which creates an electrical current, which powers your computer, guitar amp, and refrigerator so you can write a brilliant musical that wins you a

---

* Rudolf Clausius and William Rankine were standouts in the budding field of thermodynamics. They were also early contributors to muttonchops and the stylish chin curtain.

Tony and then open an icy cold bottle of champagne to celebrate. Thank you, first law of thermodynamics! Still, no new energy was created. Potential energy in coal, natural gas, or nuclear material was converted to heat energy, then to motion (kinetic energy), which was converted to electrical potential and sent out on power lines to your computer, guitar amp, and refrigerator. The only new creation was your brilliant musical.

If we didn't extract the natural gas and convert it to heat in the power plant, it would still be quietly hanging around underground with its potential untapped, waiting to be discovered like a gifted actor doing community theater in Albuquerque.

Potential energy is eager for its turn to do something exciting. There are a few different kinds of potential energy. One of the easiest kinds of potential energy to visualize is gravitational. A snowball ready to fall down a mountain has plenty of this. That snowball can convert its elevation potential energy into a little avalanche by simply falling. Another kind of potential energy is chemical. The gas in your car is a good example of this. Because of its particular chemical makeup, it can combust and release energy that revs your motor.

My favorite kind of potential energy is elastic. (We all have a favorite kind, right?) Elastic potential energy is a scientific way of describing things that when stretched, pulled, or otherwise deformed readily spring back into shape. A bow and arrow, a catapult, a rubber band, and a spring can all be forced out of their most comfortable position. When they bounce back into their resting position and spend their elastic potential energy, arrows fly, rocks sail over castle walls, and the trusty clothespin snaps shut.

To see every possible form of energy conversion, simply watch a few episodes of the old cartoon series *Road Runner*. Wile E. Coyote uses elastic, gravitational, and chemical energy inventively and unsuccessfully over and over again. (He's a talented engineer, but needs a good project manager to help him with timing, execution, and a roadrunner behavior study.)

## Kinetic Energy: Speeding Cannonball

The best way to experience the conversion of potential energy to kinetic energy is to go cliff diving in Jamaica. Step one: climb the cliff. As you climb, you are gaining potential energy in the form of elevation. Your body weight multiplied by the elevation above the surface of the water is your newly acquired stash of potential energy. (You burned chemical potential energy in the form of calories to earn that elevation—remember, you only convert energy; you don't create it.)

When you get to the top and stand on the cliff looking down to the water, you contemplate converting all your stored potential energy into kinetic energy with a jump. You are also contemplating how likely your swimsuit is to stay on your body when you make that big splash. It might be easier to answer that last question if you knew how much kinetic energy you will have when you hit the water. So you engage in a little conversion calculation while you stand at the cliff working up the courage to jump.

Your potential energy is simply your weight multiplied by the height of your cliff position, so take your 140 pounds and multiply it by the 20 feet you are about to jump. (Hey, no one is weighing you in at the cliff before you jump—you can tell this story any way you want.)

You know you are poised and ready to take 2,800 foot-pounds of pure potential energy and convert it to speed, adrenaline, and a tiny nervous accident in your swimsuit. When you step away from the cliff and into thin air, your body will drop straight down and start speeding up. It doesn't matter if you jump in a gorgeous, silent swan dive or a screaming windmill. Every foot you drop, you are converting height to speed or, more scientifically stated, potential to kinetic energy. When you hit the water, you will have converted all your potential energy into kinetic energy.

What is 2,800 foot-pounds? It's the force required to move 2,800 pounds 1 foot, or 1 pound 2,800 feet. That's not very helpful to you as you stand there wondering if there are sharks watching from below the water, trying to stifle their giggles as they wait for you to enter their food court. What will 2,800 foot-pounds feel like in the form of a belly flop? You decide to figure out how fast you'll be going when you hit. That's one's easy—put potential energy on one side of an equation and kinetic energy on the other:

Potential energy is mass × gravitational acceleration x height

Kinetic energy is ½ x mass × velocity$^2$ for kinetic energy

If all your potential energy is converted to kinetic energy,
the equation is: $mgh = 1/2mv^2$

Your mass cancels out on each side (see, it really didn't
matter how much you said you weighed). Then you plug
in the numbers you know and solve for velocity.

$32.2 \text{ ft/s}^2 \times 20 \text{ ft} = \frac{1}{2} \times v^2$

You got it from here, right? Multiply the numbers on the left side and then multiply each side by 2. Finally, take the square root of each side to get velocity by itself on the right. You find you will hit the water at about 36 feet per second. That's about 24 miles an hour, which is . . . wow, that's fast to be flying through the air mostly naked. But remember, you're hitting *water* at 24 miles an hour, not a brick wall. You consider the advantage of water slowing you down gradually instead of suddenly. The water will take a second or two to stop you after you hit the ocean's surface, so you have that time to spend all the kinetic energy you'll have on moving the water around. If you were hitting pavement, it wouldn't move around to make room for you like water will. All your kinetic energy would be converted into breaking your legs. Of course, if you were jumping onto pavement, there would be no chance of landing into a circle of hungry sharks. There are pros and cons to each, certainly, but a pink water-slapped stomach followed by your fastest swim to shore are better than a body brace.

By now you have been standing at the cliff long enough that you're starting to get some serious tan lines. You have scribbled calculations in the sand at your feet and locals are starting to gather to see if you need help. So you jump. You convert your potential energy in the form of weight and elevation into kinetic energy in the form of weight and speed. You gain a wonderful life experience, the thrill of scientific experimentation, and a video of yourself making a very weird terrified yodel before getting a brutal water wedgie. All well worth it.

## The Physics of Life: Following the Strict No-Spin Rule

When a car gets stuck in mud or snow, it takes perfectly good potential energy in the form of gasoline and converts it into the kinetic energy of spinning wheels, flying mud, and an overheated engine. The driver takes potential energy in the form of calories and converts them to the kinetic energy of swearing and punching the dashboard. The results are not productive, valuable, or repeatable in polite company.

We can learn a lot from a limousine stuck in a New Year's Eve snowstorm or a monster truck wedged into a more monstrous ditch. I decided a few years ago to never spin my wheels again. I listed what I considered wheel-spinning activity: worrying, fussing, complaining, and describing how busy I was to someone who was inviting me to a party that I didn't have time to attend. I would not do any of those anymore. If I needed to rest, I'd rest. If I needed to work, I'd work. If I couldn't attend the party, I would politely decline and send flowers. I would not spin my wheels.

In my track and cross-country racing days, we called training that is neither very challenging nor very restful "gray zone training." A runner's body improves fastest if they alternate between workouts that are at a gut-busting race pace and a true rest pace. If they grind through the same mediocre pace every day, a runner's body doesn't feel the need to develop more lung capacity or muscle. Also, the runner risks getting injured or burned out because they are never truly resting.

Despite understanding the "gray zone" concept in running, I was violating it in my work life and creative projects. I did a lot of fussing and spinning when I should have been either resting or in full productive kinetic motion.

So I tried something new. When I noticed myself spinning my wheels, I changed activities. I committed to doing something completely useless, such as watching a movie with a talking animal as the main character or making cupcakes—the easy kind, from a boxed mix, nothing tricky. When my head was clear, I'd get back to work. I found that if I stopped spinning my wheels to truly rest, I was ready for action again very quickly and I was much more efficient after the pure rest of cartoons and chocolate frosting.

We have only so many years in our lives, so many hours in the day, and so much creative capital to spend. If we want to spend it efficiently, that requires a bit of rest. If we spin in constant motion our lives are useless smoke, noise, and burning rubber. Personally, I don't want to spend a second spinning because before I'm done with this life, I want to spend every bit of my potential. I want to put it in motion and make kinetic any talent, luck, strength, and humor I was given at birth or earned through experience. I intend to convert it all to action so there is no wisp of energy left to float out with my final breath. Nothing even for the quickest savior to catch in his hand and push back into my chest.

## ((( PHYSICS PRACTICE )))

**1. Which one of these does not have potential energy? Coal, crude oil, a coiled spring, or a swinging pendulum at its low point?**

ANSWER: A pendulum as it swings through its low point doesn't have potential energy. It has kinetic energy. It's moving fast and will swing up to stop at its maximum height (and maximum potential energy), then back down to its maximum speed (and maximum kinetic energy.) So, at its low point, a pendulum is chock full of kinetic energy but not potential energy. Coal, oil, and a coiled spring all have potential energy to spend.

## 《 TRY THIS! 》

1. Sneak into a gymnastics training gym and get on the trampoline. Jump as high as you can while shouting tips to the gymnasts.

   A. When is your kinetic energy the highest? When is it the lowest?

   B. When is your potential energy from elevation the highest? When is it the lowest?

   C. When is your elastic potential energy the highest? When is it the lowest?

   D. Do you really have any useful tips for the gymnasts?

ANSWERS:

   A. Your kinetic energy is the highest when you are going the fastest. That is right when you leave the trampoline heading up and right before you touch it again on the way back down (just like a ball that's thrown in the air). Your kinetic energy is the lowest at the very top of your jump, when your velocity is 0, and at the very bottom of your jump, when the trampoline is stretched down and ready to throw you back into the air.

   B. Your potential energy from height is the greatest when you are at the top of your jump. It is the lowest at the very bottom of your jump.

C. Your elastic potential energy from the deflection of the trampoline is highest when you are at the bottom of your jump and the trampoline is dipped down with the force of your landing and ready to toss you back up in the air. It is the lowest whenever you are in the air. The trampoline has done its pushing on you and hangs out waiting for you to drop on it again.

D. Well, no. But everyone likes to hear the occasional "You got this!" Sure, trespassing in a private gym and using their equipment may be a crime, but being supportive is not.

# 5

# KNOW YOUR TYPE
## THE CHEMISTRY OF ATTRACTION AND BONDING

In what I now recognize as a brilliant strategy to motivate girls to take more than the minimum required science and math in high school, certain classes were offered only at the boys' school. If we wanted to take chemistry, calculus, or physics, we needed to cross the street and enter the Guy Zone. My classmates and I became intensely interested in furthering our math and science studies. Our surprised parents listened as we explained that these courses were imperative to our future success as doctors, astronauts, and celebrity hair colorists.

On our first day of chemistry class, we walked over in a tight pack of plaid. We instinctively knew that a strategically timed point-and-giggle from our knee-socked herd could neutralize any aggressor, possibly for the remainder of his high school career.

In the classroom we stuck together, taking over the rows of

desks closest to the windows. The boys shuffled in and took the rows closest to the door. Then, a giant man in a pressed button-up shirt and khaki pants breezed through the door with his arms open wide and announced, "This always happens! The girls on one side of the room and the boys on the other!"

He stood in front of one of the boys' rows and ordered, "Up, up, up, all of you in this row." He then stood in front of the row of girls next to me. "Good afternoon, ladies, welcome to our school. If you would please be so kind as to move into this row." He pointed to the row that was currently being vacated by the cluster of confused boys. In this way, he arranged us in alternating rows of boys and girls.

We looked desperately at each other. There was chaos, the sound of dropping books, the smell of cotton candy lip gloss as the girls in the forced migration managed a brave last touch-up. We regarded our new teacher with mute horror. What else was this monster capable of?

He turned to the board and wrote his name in chalk. It was impossible to pronounce. He then turned back to us and said, "You may call me Mr. G. Ladies in this row, your lab partner is the gentleman in the row to your left." He took a step to his right and repeated the process until we were all paired off.

"Much better. The junior prom is in a few months. How will you ever find dates if you don't *speak* to one another?"

The boy across the aisle from me was a disaster. Baggy pants, multiple cowlicks, smelling of gasoline and tater tots. Did I have to go to the junior prom with *him*?

"Take a moment to introduce yourself to your lab partner."

"My name is Christy," I said to Gas Tots, checking out the stripes he'd colored on his tennis shoe soles with a blue pen.

"I'm Spanky," he replied and gave me a big, goofy smile. I wanted to ask for his birth certificate, as I doubted very much his parents named him Spanky. Instead I extended my hand and shook his. It was the Christian thing to do.

As the weeks passed and Mr. G immersed us in his world of chemistry, I began to understand him better. He saw everything in terms of bonding, coupling, combustion, and attraction. He couldn't resist the urge to pair us up and encourage us to bond. Our class became more comfortable with our forced marriages while Mr. G helped us crack the code of the periodic table: One hundred and eighteen colored boxes with one or two letters inside and a number in the upper left corner of each. These boxes were lined up in eighteen columns, seven rows deep, with boxes missing on the top and some extras tacked on at the bottom.

It became clear that all the answers to his questions were on it. Like a game show host, he would ask, "Who can tell me how many protons carbon has?"

We'd scan the periodic table to find the box with a C in it, look at the number above it, and several of us would shout, "Six!"

"You people are brilliant. You scare me," he would say with a straight face.

Mr. G built our confidence until we understood the basics of the periodic table of the elements. Each box with letters in it represented a chemical element, a particular type of atom. There is a box for hydrogen, one for gold, one for neon, one for all of the known elements. In each box are the letters representing that element. Carbon is very sensibly represented by the letter C. The number above carbon's C is 6. That is the number of protons it has at its center. The number of neutrons in the nucleus of the atom can be equal to the number of protons. (If it isn't, that atom is called an isotope of the atom and things can get a lot more interesting. For the moment, we were just looking at atoms with an equal number of protons and neutrons.)

Carbon, with its 6 protons and 6 neutrons clumped together in its nucleus, had 6 electrons spinning around it. We could think of it like a little solar system.* Mr. G pointed out the window to the football field and gave us some perspective on the atom. "If the nucleus were the size of a marble, the electrons would be flying around in a space the size of that whole football field." I pictured the electrons zipping in laps around the track in tiny spiked racing shoes.

He spread his arms out and said, "The world is made of atoms, and atoms are made of mostly nothing."

It almost sounded like Mr. G was implying that God hadn't done too much after finding the world "without form and void." I wondered if my sophomore Old Testament teacher, Sister Paula, knew about God phoning it in like this. Very disappointing. Honestly, God. We expect better of You.

With little positive and negative signs on the chalkboard, Mr. G showed us that protons have a positive charge and neutrons don't have any charge, so carbon's nucleus has a positive charge of 6. This positive charge of 6 compels 6 electrons, each with a negative charge, to spin around carbon's nucleus. For the moment, each atom was a tiny, balanced universe with the same number of protons and neutrons in the nucleus, and the electrons hovering around it in perfect balance.

A week into our exploration of the periodic table, Mr. G

---

* In physics class in our senior year, we'd hear some very strange stuff about electrons that would blow away our solar system model of the atom.

explained that we were now masters of the chemical universe. We knew how to find the number of protons, neutrons, and electrons in each atom, and these defined everything about that atom: its melting point, freezing point, weight, hardness, density, ability to transmit electrical current, and its bonding behavior with other atoms. "Let's talk about that bonding," Mr. G said. "Why are the atoms so attracted to each other?"

When Mr. G said "attracted to each other," did his voice drop into a lower register or was I just hearing it that way? Was I the only one picturing the atoms having sex? Mr. G explained how atoms wanted full outer orbitals of electrons. Oh, yes they did. They wanted it badly. (Cue the make-out music.)

Well, thank goodness atoms do feel compelled to *bond* with each other. When two hydrogen atoms bond with an oxygen atom, we get water. When sodium and chlorine mate, we get salt. The proteins that we ingest are made of amino acids, which are ordered clumps of nitrogen, hydrogen, carbon, and oxygen atoms. Atoms on their own are great, but we really need them to get together and create the molecules and compounds we need for growing, breathing, and properly seasoning a bowl of popcorn.

## The Atomic Dating Game

As we focused on the desires of carbon, oxygen, hydrogen, and other atoms in the top rows of the periodic table, we learned how they got their full outer orbital of electrons. These outer orbital electrons are called valence electrons. The number of valence electrons defines an atom's neediness. For most of the atoms we were interested in at the moment, the magic number was 8. With less than 8 electrons in their outer orbital, these atoms are lonely and unfulfilled. They want to bond. They bump their eager outer orbitals together and share valence electrons.

Mr. G explained how these electrons were arranged in the atom, and the term "outer orbital" started to make sense. It was like their skin. Atoms with higher numbers on the periodic table have more protons and neutrons in their nucleus. This means they need more electrons in the orbitals around them to have a balanced charge. The electrons tend to follow the same pattern in every kind of atom. The first 2 electrons orbit closest to the nucleus. Then the next 8 fill up the next orbital, the next 8 fill up the orbital after that, and on and on.

Argon, with an atomic number of 18, has 2 electrons in the orbit closest to the nucleus. In the next shell out, there are 8 electrons. In the third shell, there are 8 more. Argon is happy and fulfilled with its complete 2-8-8 electron setup and needs no more electrons. Chlorine, just to the left of argon on the periodic table, has an atomic number of 17. That means it has 17 protons in its nucleus with 17 electrons hovering around it. This gives chlorine 2-8-7 electrons in its inner to outer orbitals. Uh-oh. That's not a full 8 on the outer orbital. Chlorine is on the prowl for another electron. Who has 1 electron in her outer shell? Lonely little hydrogen, that's who—sitting on her porch in a sundress dreaming of electrons. When chlorine pulls up in his dusty Camaro, it is *on*.

I pictured hydrogen wrapping her teensy electron orbital legs around chlorine as she willingly shared her 1 outer electron with chlorine's 7. They were making each other so happy, completing each other's atomic lives in a committed 8-valence-electron relationship. This was called covalent bonding: atoms sharing electrons to form a full outer shell, exchanging household appliances on Valentine's Day, and saving up for a second honeymoon to Mount Rushmore. As we matched atoms up in covalent bonds, the columns of the periodic table, called groups, took on bonding personalities.

On the far left of the periodic table, on the top of the first column sits Group 1. Hydrogen and her left-column friends below are shiny and soft. With only 1 electron zipping around their outer orbitals, they don't like to be alone. This makes them somewhat (ahem) indiscriminate about bonding. They just want to be loved. They are vulnerable and easily unhinged, the Blanche Duboises of the atomic world.

The second column in the periodic table, Group 2, contains the perpetually overlooked earth metals. With two available electrons in their outer orbitals, they are just as willing to share their shells as Group 1. The atoms in Group 2 try very hard to marry up to, say, oxygen, to get a full outer shell of electrons.

Columns 3 through 12 are short columns called the transition metals, but who are we kidding? All we really care about are the precious metals in column 11—copper, silver, and gold. They are the stars of the metals: malleable but strong, beautiful and useful. They are mined and separated from their duller atomic cousins because they are best when they are pure. These beautiful creatures tend to outshine everyone around them; they can't help it. They are tarnish-resistant and difficult to counterfeit. They have an electron or maybe two to share, depending on who is asking and how they state their question. They are maddening and evasive and we love them for it.

Groups in columns 13, 14, and 15 have team captains of boron, carbon, and nitrogen. They make wonderful covalent bonds. They know how to create partnerships, families, and organize a themed brunch. The Boron Family has 3 valence electrons to share, the Carbon Clan has 4, and Team Nitrogen has 5. They need only a few more electrons to make 8. None of the atoms in these sororities is needy. They are dignified members of the atomic community, not like those sloppy train wrecks in columns 1 and 2.

Groups 16 and 17, with oxygen and fluorine on the top, have 6 and 7 electrons in their outer shells, respectively. They bring a lot to an atomic relationship, and they know it. This gives them confidence and a sense of entitlement. They can be generous about sharing their electrons, but also explosive and demanding.

On the very far right of the periodic table you will finally find a column of atoms that doesn't need any electrons—the noble gases. They have all they need in their outer shell already, a full house of 8 valence electrons. They don't care to bond. They keep to themselves. The other atoms consider the noble gases stuck up, but the noble gases don't care. Neon, argon, and the others on the right side of the periodic table are the Emily Dickinsons and

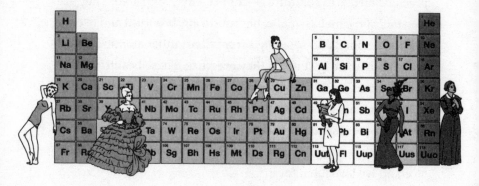

Nikola Teslas of the atomic world: brilliant, productive loners. Maybe they aren't the greatest dancers, but they always keep their pants on at dinner, which is more than we can say about lithium and calcium.

## Magnetic Attraction

After we had a solid understanding of covalent bonding, Mr. G explained ionic bonding. This is when the coupling truly started to get steamy and reckless. In ionic bonding, the atoms don't share electrons as nicely as those covalently bonded atomic couples. Instead, they rip them from each other in a fit of hungry passion. After one atom has lost electrons and another has gained those electrons, they are oppositely charged—stuck together with nothing in common but a crazy magnetic attraction.

Mr. G again described chlorine with 7 electrons in his outer orbital hanging out on the right side of the periodic table. This time, though, it didn't partner with hydrogen in a covalent marriage. His date was with sodium. With only 1 electron in her outer orbital, she looks around impatiently in her column on the left side of the table. Oh, sure, she loves her 1-electron-outer-orbital buddies lithium and hydrogen, but they can't help her. She needs an atom with 7 electrons to share to make a full 8. Sodium sneaks across the periodic table in the middle of the night, ignoring the crude remarks from the base metals, and falls into the arms of chlorine, giving up her only spare electron to him. But he doesn't share it. No, he hogs that 1 electron to fill his outer shell. Without her only outer electron, she becomes positively charged and he becomes negatively charged and they are pulled together with a magnetic force resulting from his selfishness. They are a doomed and passionate molecule couple, not entirely fulfilled, but unable to leave.

Ionic bonding wasn't like electron-sharing, respectful covalent bonding. But at least the atoms involved only had electron eyes for each other. Mr. G was about to tell us about some atoms that were not monogamous. Not even close.

This obsession for 8 outer orbital electrons leads atoms to do some things that would horrify their atom parents. Mr. G. described sulfur swinging each of its smelly arms over a hydrogen atom to make a threesome; metallic atoms bonding in long, indecent chains; a ring of carbon atoms creating an incestuous benzene ring. It was downright obscene. These atoms obviously had not taken ninth-grade health class with a "biblical focus."

As we explored the R-rated personal lives of atoms, Mr. G's chemistry experiment with his rows of boys and girls was similarly heating up. Desks inched across the aisles; heads leaned over Erlenmeyer flasks close enough for safety glasses to knock awkwardly against each other. Mr. G's class was bubbling over.

My lab partner, Spanky, had an outer shell that was of no interest to me. I wanted to bond with lanky, wisecracking, smartypants Patrick. He sat behind Spanky, one leg stretched out in the aisle, expertly directing his voice in such a way that apparently only I could hear him.

While Mr. G demonstrated the unchanging temperature of boiling water and leaned toward the Bunsen burner, Patrick narrated in a whisper, "If he gets any closer to that thing, his fur brow is going to go up in flames."

I continued to look straight ahead, but Patrick knew I was listening. "Okay, here's the plan. I'll tackle him and smother the brow fire before it spreads. You hit the gas valve before the room blows. We'll be heroes."

It was not easy to maintain a straight face, but I did so as I turned to give Patrick a look that attempted to express many things at once: *I am trying to learn chemistry here, please stop distracting me. Also, I'm wondering if you have chest hair.*

Over the next couple weeks, Spanky scooted up the aisle to partner with Denise, who, like him, had blue ballpoint pen stripes on her shoes and smelled like cafeteria food. Patrick took Spanky's seat, and by the time we were using eyedroppers to add bases to acids, Patrick and I were lab partners. We almost touched knees as we dipped litmus paper into vinegar and watched the color change.

Patrick compared the orange strip to a color wheel and asked, "Does it look more like color number 2 or color number 3?"

"They're both orange," I replied. "Are we supposed to be able to tell the difference?"

What I really meant was, *I want to pin you down and punch you in the face. Then I want to kiss you. Sometimes I have confusing dreams about you.*

## The Physics of Life: Know Your Atomic Identity

I thought Patrick might ask me to the prom. In retrospect, I can see why he didn't. When I saw him in chemistry class, I was focused on calculating a molar mass or finding a melting point.

When I saw him at track meets, I was concentrating on what my 400-meter splits needed to be that day in my race. I didn't know then that before someone asks you out, you might need to actually look up from what you're doing and smile, preferably directly at them. Otherwise, they may not know that you like them.

My behavior was typical of a noble gas. I didn't know at the time that was my atomic dating identity. Those noble gases on the right side of the table with their full outer orbitals and mysterious superhero names (helium, neon, argon, krypton, xenon, and radon) don't react with other atoms. They are not terribly flammable. They are earnest and hardworking. They have little interest in bonding with other atoms. Because their outer orbitals are full, the noble gases are contented loners.

I don't flatter myself by thinking I'm anything like the rare and expensive xenon or krypton. I consider myself simply a decent laboratory-grade argon. I wish I would have known this in high school. It took me years to figure out my atomic identity, probably because so much attention is paid to the bonding types. Our culture is obsessed with bonding. Movies, songs, and music videos celebrate first kisses, dance parties, team sports, true love, and other outer orbital-filling activities. Rarely are music videos made about getting up early on a Saturday to go running alone or staying late at work to compare turbine efficiencies.

If you recognize your atomic identity and don't like it, you might think you'll just lose or add a proton and jump over to another spot on the periodic table. Oh, if it were only that easy. That's exactly how the first chemists tried to get rich—turning one element into another. These entrepreneurial types learned this chemistry lesson the hard way.

## Alchemy Is Tricky

Those early chemists noticed that dumpy lead had a lot in common with sexy gold. They were both malleable and didn't corrode. With the modern advantage of the periodic table, we can see why lead and gold seem like sisters. They are on the same row of the table. Lead has 82 protons in her nucleus while gold has 79. Rip 3 protons out of lead and you have gold. Easy! Now we can make a giant gold tiara!

Unfortunately, like pole vaulting or the Argentine tango, making gold from lead is not as easy as it looks. An impressive amount of energy is needed to pull protons from or add protons to a nucleus. Fission (splitting the nucleus of the atom into smaller nuclei) occurs naturally in some cases. But unless it's happening on its own, fission and its opposite, fusion (adding protons to the nucleus), can't be accomplished with a handheld blowtorch in your garage and a casual knowledge of atomic structure. If you are determined to mess with numbers of protons in an atom, you need to know what you're doing because you have wandered into the territory of the Manhattan Project. Without access to the equivalent of a nuclear reactor or atom bomb, no matter how much you want to switch up numbers of protons in a nucleus, it won't happen.

It must have been exciting for those early chemists when they realized there was an orderly way to line up all the atoms in a table. If they laid them out by weight in columns of 8, a simple pattern of properties and bonding behavior emerged. Later, when elements were discovered that didn't fit, the periodic table grew to adapt. When there was a gap in the table, scientists knew the element existed in the universe somewhere even before they could physically find it. The periodic table doesn't have blank boxes. It may have an element that appears only fleetingly in a lab, but no blanks. We need every element for the periodic table to make sense, for the universe to have all its building blocks.

Whatever we are, it is right. Every bonding type has a place and a purpose in the world, even if it is, like argon's, to float alone in the atmosphere while the carbon bonders below enjoy a perfect day of birthday parties and garage sales. I have now accepted that I am a noble gas. Before I did that, though, I tired myself out trying to be wild and carefree hydrogen. It was exhausting. I couldn't stand waking up to dirty wineglasses in the shower and confetti in my bra. It wasn't me. The same thing would happen if any of my hydrogen friends tried to be a noble gas. They would die of boredom spending a day looking at old maps in the library and then heading straight to the gym for some lap swimming.

If you are loyal, bondable bromine with 35 protons in your nucleus and 35 electrons in orbitals around it, you might be tempted to add a proton to become the exotic loner krypton. Before attempting any such changes, however, it is crucial to ask yourself if making the switch is worth the mushroom cloud and radioactive fallout. Changing careers and adding definition to your triceps is fine. Changing who you are at your very nucleus—how you bond, live, and love—is more complex.

If you find yourself saying to a friend, "I don't know why, but I just like unavailable men," or "the girls I date eventually steal all my furniture while I'm at work," maybe you haven't yet found your true bonding identity. I know there are psychiatrists and family counselors who will say it's not that simple, but isn't figuring out who you really are and how you like to bond a good place to start?

Like atoms, at our core we are something so distinct that to disassemble our core is nuclear fission or fusion—terribly energy-consuming, explosive, and likely to leave dangerous leftover parts. If you are dependable iron and you try to be heady hydrogen or exotic einsteinium, you will likely make a mess of yourself. It will hurt. And it won't work. Be an iron person if you are iron. Or be carbon if that's how you're made. There's a place for every type of bonder in the world, just as there is a box for every element in the periodic table. You are somebody's match. Or you are a noble gas. Either way, you have a place in the universe.

## ((( PHYSICS PRACTICE )))

1. You are hosting a cocktail party. Your guests are 1 atom each of chlorine, magnesium, carbon, oxygen, sodium, arsenic, and argon. Things get off to a rocky start. Chlorine (7 outer electrons) is hitting on magnesium (2 outer electrons) in search of additional electrons. This should be easy for chlorine, but when they get close, their 9 combined electrons make them both feel overly warm. They sweat nervously and look around the room for a way to shed their extra electron. Carbon (4 outer electrons) and oxygen (6 outer electrons) start off well enough. They snack on celery sticks and cheese cubes while they chat about the best place to hire a birthday clown and how grateful they are for electronic baby registries. They consider each other equals since they aren't those low-electron unfortunates. As they get closer and form carbon

monoxide, though, they have 10 outer electrons, making them over-charged. Sodium (1 electron) is getting drunk in the corner while arsenic (5 electrons in his outer shell) refills sodium's glass and asks if she has two friends she can call. Argon (full outer shell) is doing the dishes in the kitchen and silently judging them all. She thinks it's sad, just *sad* the way these atoms behave at parties.

Cool, mysterious krypton shows up with another carbon. What can you do to get your party back on track?

ANSWER: Tell oxygen you need his help opening a bottle of champagne in the kitchen. Ask magnesium if she'll come and taste the salsa you are making to see if it's spicy enough. While magnesium and oxygen get to know each other in the kitchen, pull the abandoned chlorine over to sodium. Ask krypton to keep an eye on arsenic while you introduce chlorine to sodium. With chlorine's 7 outer electrons and sodium's 1 free electron, the two will quickly fall into a salty embrace. Arsenic will be angry because he thought he was making progress with sodium, but with krypton staring him down, arsenic won't break any of your wineglasses. The two carbons need no introduction. They will immediately start making plans to start a nonprofit that provides social media consultation to underprivileged children. Ask argon to drive arsenic home before he has any more drinks and makes a scene. Argon will be looking for a reason to go home early anyway. And now that you have krypton to help you keep the peace, you are in good shape. Side note: Don't make a fool of yourself trying to flirt with krypton after everyone leaves. He's not interested.

2. In which group of elements do you belong? Based on this, what kind of person should you date or marry?

**ANSWER:** If you are a noble gas, only date other noble gases. Otherwise you will be a disappointment and heartbreak to a regular bonder. If you are in the carbon or oxygen columns, try to stick with someone in your own column. An equal. Don't be tempted to date a needy left-column type. If you are an eager bonder with few valence electrons on the left side of the table, get a grip. Don't throw yourself at an oxygen type. Make sure they appreciate you first and understand that the few valence electrons you bring to the relationship are worthwhile. Don't bond with other needy left-table types. You'll still be short electrons and be looking for something more fulfilling. And for goodness sake, cover up. You're not going out in that, are you?

# 6

# DON'T ANSWER THE DOOR IN YOUR UNDERWEAR

## THE IDEAL GAS LAW

The ideal gas law is a simple little equation that describes how volume, pressure, and temperature are related in most gases and in most circumstances. The lawyerly language at the end of the last sentence is meant to remind us that this law can't be used everywhere in the universe and all the time, but for the situations we run into in our daily lives, this law is a real gem. Also, the ideal gas law is a great model to show us how variables affect each other in a closed system. It can be tweaked slightly to create the ideal task law—a guide for performing under pressure.

If you are a math person, you will think of the ideal gas law as $PV = nRT$ where P is pressure, V is volume, n is number of atoms or molecules of gas we are talking about, R is a fancy constant with units designed to make sure the other numbers play nice together, and T is temperature. If you aren't a math person, you can simply think of the ideal gas law as a strong contender for the most obvious scientific law ever. The ideal gas law and its simple barefoot cousins Charles's law and Boyle's law are ways to calculate and quantify things you already know intuitively. For example, if we let air out of a balloon, the balloon will get smaller—as long as we don't get cute and start changing the temperature or the air pressure on the outside of the balloon. Most of us were aware of this shrinking balloon phenomenon before first grade. This is not breaking news.

Despite not being exotic, the ideal gas law does invite us to examine all things gassy in more detail. It allows us into the secret lives of air molecules. They don't just quietly hover in one spot. (When I say molecules, I mean both molecular gases such as $CO_2$—carbon dioxide—or plain old atomic gases such as helium; they can be considered the same creature in our ideal gas world.) If they were big enough to see, we'd notice gas molecules dancing around, always in motion. The more energy they have, the faster they move. As they gain energy and move faster, the temperature rises. The only time molecules stay almost perfectly still is at a temperature called absolute zero (minus 459 degrees Fahrenheit). Otherwise, those molecules are dancing fools.

To illustrate the activity of gas on a molecular level, Mr. G put two beakers of water on his desk. One was icy and one was boiling. He took out a metal bulb that looked like an antique Christmas ornament with a piece of thin pipe connecting it to a small valve and pressure gauge. He plunged the bulb into the icy water. The gauge reading the pressure inside the bulb didn't do anything. "Fascinating," whispered Cute Patrick behind me. Not a very memorable experiment so far.

But when Mr. G put the bulb in the hot water, the pressure shot up in the constrained volume of the bulb. He explained that the molecules inside the bulb were moving around more since they were heated up. Moving molecules smack into each other, and push on the walls of the metal bulb. This smacking of many molecules translates into the measureable pressure.

Mr. G opened the valve to let a little hot air out, so that the pressure was evened out with the atmospheric pressure in our classroom; then he closed the valve. When Mr. G dropped the bulb in the cold water again, the thin metal bulb crinkled and caved in like an empty beer can under a cowboy boot. As Mr. G explained what had just happened, I envisioned the dancing molecules inside the bulb.

Originally, the room temperature molecules in the bulb danced with each other in a formal minuet as they listened to Mozart. They smiled politely and allowed only their little molecules' hands to touch. When the bulb was dipped in the hot water, energy was added (heat), and the molecule band started to play some early rock 'n' roll. The boy molecules tore off their stiff coats, the girl molecules let their hair down, and they started to swing dance, bumping into each other more than when they were stiffly pacing to the Mozart tune. The show-off molecules threw their partners in the air and slid them through their legs. All this bumping into each other and the walls drove up the pressure inside the metal bulb. As the temperature continued to rise, the band picked up the tempo even more and the molecules started dancing in a frenzied mosh pit. The atoms threw off their cardigan sweaters. They ripped and spray-painted the T-shirts that hung on their sweaty little molecule bodies. They gave each other crooked mohawks and lip piercings while slamming away. That's what happens when molecules have a lot of energy and nowhere constructive to use it.

With the molecules tightly contained and heated up, the pressure inside increased. Those air molecules were dancing away, slamming into the sides of the bulb. They busted out their best moves, but dammit, they kept slamming into those walls. They needed space to express themselves! When Mr. G let some of the crowded molecules out of the bottle to return it to the pressure it was before he heated it up, the molecules had room to spread out. They were dancing as furiously as before, but there was less crashing into walls and more freedom to cut loose.

When the air on the inside of the bulb was cooled back down again, the energy dropped, the music slowed down, and the molecules became self-conscious. They realized the dance floor was half empty. Where had everyone gone? The molecules cleared their throats and looked for their discarded clothing. They danced politely again, staring at their shoes, awkwardly avoiding eye contact. They no longer needed as much space. And now that there were fewer of them, there was very little bumping against the inside of the bottle. The pressure inside was lower than the room pressure outside of the metal bulb, and the walls caved in. The party was definitely over.

Years after watching Mr. G's metal bulb demonstration, I returned to my truck on a hot day to find that a can of root beer had exploded in the passenger's seat. I knew why. When the carbon dioxide bubbles in the root beer heated up, they put pressure on the inside of the can. Eventually that pressure was great enough to demand more volume. Since the can couldn't gently expand like a balloon or a bike tire, it burst dramatically and sprayed the cab of my truck with sticky root beer. My truck had the pleasant smell of an old-fashioned ice cream shop after that.

(If the root beer had frozen on a cold day, the can also would have exploded. More on that later. Honestly, you just can't win with these cans of root beer.)

## Avogadro's Impressive Number

In the ideal gas law, the little n represents the number of gas atoms or molecules involved. Talking about numbers of atoms, though, is problematic. The numbers get ridiculously large in a hurry. So, in the same way that you might order six dozen doughnuts instead of asking for seventy-two doughnuts,* we describe quantities of atoms in moles. A dozen is twelve, and a mole is 6.022 and $10^{23}$. It's a big number, but it's still just a number with no units, such as pounds or gallons. Just $6.022 \times 10^{23}$. Like the names Madonna, Beyoncé, or Voltaire, nothing else is needed.

That particular number was chosen because it's the number of atoms in a gram of carbon-12, the most common form of carbon, with 6 protons and 6 neutrons. But how we got to the number isn't crucial. Using moles is convenient when talking about a *lot* of atoms. For example, if you read the pressure meter on a tank of pure oxygen and knew the volume and temperature of the tank, it would be simpler to report that there were 3 moles of oxygen in the tank than to write that there are 1,806,600,000,000,000,000,000,000 atoms of oxygen in the tank.

About fifty years after Italian physicist Amedeo Avogadro's

---

* Whoa. Take it easy on the doughnuts!

studious life ended, this large number was called Avogadro's number to honor his work studying the relationship between the mass and volume of gases. He was a pretty serious guy, but I think he would be privately tickled to know that high school students are memorizing and mispronouncing his name to this day. (The stresses are on the same syllables as "avocado." If you will now forever associate the ideal gas law with guacamole, you are not alone.)

## The Ideal Emergency Plan

You'll be really glad you understand the relationships among pressure, volume, and temperature of a gas if you are on a desert archaeology dig and your radio dies. As the archaeology undergraduates begin to panic as the sun goes down and the temperature drops, you will already be formulating a plan to signal your location to base camp. You have matches, but there is no firewood in sight. You don't have much time before it becomes, as they say in scientific circles, colder than a well digger's rear.

You direct your handful of student helpers to each get a plastic bag from your artifact-collecting gear. This is the first step of your plan, and it forces them to get busy and stop listening for packs of coyotes while silently weeping. You then ask them to take apart one of the wire screens they've been using to sift through the desert sand. You help them make wire rings a little bigger than their hand with outstretched fingers. Then while they add some wires across each of their rings like spokes of a bicycle wheel, you tear tiny strips of your T-shirt off (in appropriate places) and use your matches to melt your beeswax lip sunscreen on them.

By now it's dark and your students are starting to size each other up in a grim *Lord of the Flies* way. You ask two of them to hold one of the plastic bag and wire contraptions by the puffy bag side and let the ring drop. You use some wire to wrap your T-shirt and lip balm candlewick to the center of a wire ring and light it. After waiting for the air in the bag to warm up and get lighter than the colder air around it, you tell the students to let go. Up goes your lovely lighted hot air balloon! You and your students release one of these elegant beacons every five minutes to signal your location until you hear a base camp jeep rumble toward you.

Saved by your knowledge of the ideal gas law! $PV = nRT$! When the temperature rises and those heated air molecules slam-dance inside the plastic bag, fewer of them can fit inside it. Because it has fewer air molecules than the surrounding air, it is lighter than the surrounding air and it rises.

## The Physics of Life: Something Has to Give

Mr. G's class wasn't the only place I learned about pressure in high school. The teachers in the rest of my classes were eager to teach me about how to handle pressure. In my high school, they loaded us up with homework. We had to write term papers in every class, even chemistry. Thankfully, I saw the prosaic possibilities in lactic acid buildup in muscles and was able to crank out a five-page masterpiece on anaerobic exercise. At the time,

I thought that each teacher didn't realize how much homework every other teacher was assigning. I assumed they were either negligent or cruel. But now I see the wisdom in their tactics. From the overload, I learned how to manage deadlines and pressure.

The ideal gas law reminds me that when one thing must stay fixed, something else has to give. Find the variable with some flexibility and work it. If you are in Scarlett O'Hara's situation and you have only an hour to make yourself an outfit from the curtains with no one to help out, keep the complexity down by going with a sleeveless maxidress with no hem on the edges. If, like Cinderella, you have an army of master tailor mice to help you with your ball gown, go for a dropped neckline and princess sleeves. When you are under pressure, decide right away what resources are available to you and what you can realistically achieve rather than panic and end up answering the door in your underwear.

## ⦅ PHYSICS PRACTICE ⦆

1. The ideal gas law states that pressure and volume are proportional to the number of gas molecules and the temperature. PV is proportional to nRT. What happens to pressure in each of these situations?

A. No more gas molecules are added and the temperature doesn't change, but the space the gas occupies is increased.

B. The volume and temperature stay the same, but the number of gas molecules increases.

C. Helium gas is swapped out for the exact same number of molecules of oxygen gas while volume and temperature are unchanged.

D. A fixed number of nitrogen atoms at a steady volume and temperature are asked by another volume of oxygen atoms to participate in drunken freeway drag racing. When the nitrogen atoms refuse because of the danger, the oxygen atoms call them Nelly Nitrogens and say they can't sit with them at lunch anymore.

ANSWERS:

A. Pressure decreases.

B. Pressure increases.

C. Pressure will be exactly the same—all gases follow the ideal gas law in the same way.

D. Peer pressure increases.

2. Put on your jaunty Italian thinking cap in honor of Avogadro!

A. How many argon atoms are there in a mole of argon?

B. How many water molecules are there in a mole of steam?

C. How many squirrels are there in a mole of squirrels?

ANSWERS:

A. There are 6.022 and $10^{23}$ argon atoms in a mole of argon.

B. There are 6.022 and $10^{23}$ water molecules in a mole of argon.

C. There are 6.022 and $10^{23}$ squirrels in a mole of squirrels. That is way too many squirrels. We are definitely going to run out of nuts.

### ((( TRY THIS! )))

1. **Get a glass vase with an opening just big enough to hold a boiled egg. Light a small piece of paper on fire and toss it in the vase, then place a peeled boiled egg on the opening of the vase. When the flame goes out, the egg will be sucked into the bottle.**

    A. **Why did that happen?**

    B. **What are you going to do now with a vase that has a boiled egg trapped inside it?**

ANSWERS:

    A. Revisiting our friend the ideal gas law (PV = nRT), we see that if we keep the number of gas molecules the same and the volume the same, pressure and temperature are directly proportional to one another. If temperature goes up, the pressure has to go up; if temperature goes down, the pressure has to follow it down. When you first placed the egg on the vase opening, the air inside the vase was hot. When the flame went out in the vase, the temperature of the air inside it dropped. What else dropped? That's right, the pressure. Now that poor egg was stuck between a low pressure pocket of gas in the vase and regular old atmospheric pressure outside the vase. Something had to give. It was Mr. Egg.

B. Yeah. Sorry about that. I hope that wasn't your favorite vase. Just kidding. You can get the egg out. With the vase right side up, put a straw into the vase, then turn the vase upside down, let the egg fall down to the opening and blow air into the vase. The air pressure in the vase will be greater than the air pressure on the outside, and the egg will pop out just as it popped in. That poor egg had a rough day.

# 7

# KNOW YOU ARE NOT ALONE
## THE EQUALITY OF GRAVITY

On the first day of Mr. Lucido's physics class he asked us to call him Coach. He enthusiastically explained that he wasn't there to teach us physics and judge how well we'd learned it. He was there instead to coach us into being physics stars. He then sent us out in pairs, armed with a stopwatches and tape measures, to determine the acceleration due to gravity.

Coach told us to test objects of varying sizes and masses. We recorded the weights of our shoes and pencils and dutifully dropped them from the hallways into the lower courtyards, yelling "Ready!" when we were poised with the stopwatch and "All clear!" when the vice principal was in perfect position to be hit in the head with a half-eaten apple dropped from our lab partner. By measuring the distance of the fall and how long the object took to hit the ground, we could calculate its acceleration. We knew the object we dropped started with a velocity of 0. Coach asked us how much speed a heavy object gained every second after we let go. What about a lighter object? After we each calculated our answers, we realized it was a trick question. Every object accelerated at the same rate, regardless of its mass. There was only one answer: 32.2. That's how many feet per second an object sped up every second it fell toward the ground.

When we protested that feathers and cats fall more slowly than rocks, Coach answered that was only true because of air resistance. A cat spreads itself out like a little hang glider when it falls to slow itself down. If there were no air, that cat could spread out all it wanted, it would fall like a dog. That's right—a *dog*. That would be one sore and humiliated cat.

Coach Lucido explained to us that Galileo confirmed the fairness of gravity by dropping different-size cannonballs from the Leaning Tower of Pisa. We found the same thing Galileo did when we dropped shoes and pencils. They all accelerated at the same rate.

In addition to being fair, gravity is single-minded. It doesn't care what's going on in any other direction. If you drop a bullet from your hand it will hit the ground at the same time as one you shoot horizontally from a pistol at the same distance from the ground. The bullet you shoot is speeding along, covering a lot of ground horizontally while it falls, but it's still falling at the same rate as the one dropped from your hand.

While Galileo is the stately godfather of physics, the first real rock star of physics is Isaac Newton. He's not just a rock star because he invented calculus and defined the laws of motion and gravity. He fulfilled other prerequisites. He had long hair and an explosive temper. Also, he was competitive, maniacally creative, and he required a very specific type of apricot tea biscuits and hard cider backstage at every one of his lectures.*

Like every legend, there are stories circulating about rock star Isaac that we can never be sure are accurate. You may have heard the one about him sitting under an apple tree. An apple falls, bonks him on the head, and he starts to ponder what exactly just happened to that apple. What force acted on it? Is that same force acting everywhere in the universe?

Whatever really happened with Isaac and the apple, we know that while any normal person would have already been making apple pie, Newton was thinking about this mysterious force more and more. He came up with the universal law of gravitation: Every object in the universe attracts every other object. That's a huge leap to make from a falling apple. Every object is pulling on *every other* object? The Earth's mass is pulling on the apple, but is mass of the apple also pulling on the Earth? Yes. That's what Newton is saying. But since the pull each object exerts is proportional to its size, the Earth doesn't respond much to the mass of the apple. The apple, on the other hand, is drawn forcefully to the Earth.

So what happens when equal masses pull on each other? Your mass is pulling on the mass of every other person in a crowded room. Why don't you all slide toward each other? Well, since you are attracted to *every* other object, and the biggest object nearby

> Every object in the universe attracts every other object.

---

* This is unconfirmed. I have not actually seen Newton's speaking engagement rider.

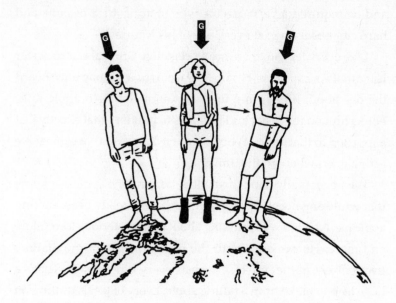

is the planet, you are all attracted to the center of the Earth in a much bigger way than you are to each other. The mass of our planet is gobnormous* compared to our teensywink† bodies or a wincy-dince‡ apple. This is why it seems like the pull toward the Earth is the only one that matters.

---

* "Gobnormous" is not yet a standard unit of measure. For reference, it is "stupid-huge" to the tenth power.

† Small.

‡ Very small.

## Gravity Mastery for Rock Stars

Gravity's insistence on treating all masses equally will help you the next time you're onstage singing and want to combine some fancy microphone-throwing moves with a stage dive.

First make sure you've mastered a quick toss of your cordless microphone from one hand to the other straight across your body with your arms outspread. You can practice in your living room. It's a nice, quick move and simple enough to execute if you keep your eyes straight ahead and rely on your peripheral vision to see each hand. A forceful toss straight across your body is all you need.

The next level of complexity is a jump straight off the stage and down to the floor. If you are still practicing in your living room, jump off the couch. Continue to throw your mic back and forth straight in front of your body while you jump straight down; you will easily catch the mic just as you did when you were standing onstage. Your body and the microphone, despite their difference in mass, will fall at exactly the same rate.

As you throw the microphone horizontally across your body,

gravity will do what it always does in the vertical direction without any thought about what's going on in the horizontal direction. It's pretty focused like that. So even if you manage to throw the mic from one hand to the other more than once in a jump, you'll catch it every time if you just do it the same way you did when you had both feet on the ground. And you will look incredibly cool doing it. So the whole routine goes like this: verse-jump-throw-throw-land-chorus-boost-from-front-row-back-on-stage-verse-chorus-big finish. Gravity and the guys in the front row have you covered.

## The Science of Sniper Fire

Experienced long-distance shooters are very familiar with the effects of gravity. They know that while their bullet whizzes toward the target, it will also be dropping the whole time just as if it were rolled off a kitchen table. For a short shot, the bullet doesn't have a lot of time to drop, so gravity can be ignored. (Don't say that too loudly; gravity doesn't like to be ignored.) But for a longer shot, gravity has time to do its acceleration thing on the bullet, and the drop starts to add up to some real vertical distance.

If you are ever part of a team rescuing aid workers being held hostage by a fussy island dictator, this understanding of gravity could come in very handy. Say, for example, you and your team have been watching the house where the hostages are being held. They are flashing Morse code messages to you with a hand mirror. They communicate that they are being treated well except for the daily hardship of trying not to laugh at the ridiculous dictator's obvious hairpiece. You are 120 yards away, ready to help them escape, but you aren't sure what they need. As evening approaches, they signal the phrase "lights out" and you realize that they need the outside lights to be turned off. You can't get close enough to flip the circuit breaker, so you will have to shoot out the big floodlight in front of the building.

Since you know you are 120 yards away and that the bullet you fire will travel about 3,000 feet/second, you quickly convert units and calculate that the bullet will be in flight for about 0.12 second. In that 0.12 second, gravity will pull down on the bullet, dropping it 2.78 inches. So you aim a few inches above the floodlight, knock it out, and the hostages make a run for it.

Since the dictator insists that his guards run in perfect synch with steady heads and stiff necks, the hostages outrun them easily and make it to safety. (Sure, you could just shoot at the guards, but that's not sporting. Being forced to run like orchestrated ostriches is punishment enough.)

## Fields of Influence: Nowhere to Hide

When we consider the effects of gravity, it's helpful to think of a gravitational field, which isn't a field per se but more of an area of influence. For example, on the surface of our planet, we clearly live in an area highly influenced by the gravitational field of the Earth's mass. Our planet's gravitational influence extends out into space— so does the Sun's and Jupiter's and that of all the other planets. Since gravitational pull gets weaker as we move away from the big mass that's pulling us, we can ignore Jupiter's gravitational influence when we are on the Earth's surface. There we are most influenced by the Earth's gravitational field and can consider only its influence when we swing dance or walk in high heels.

When a man walks into a room in a cloud of his heavy cologne, I am reminded of a field and its influence. The closer you are to the man (the source of the cologne) the more you smell his bergamot and juniper blend. If you walk away from him, you will still be in his cologne field, but you'll smell only a few high notes of citrus. If everyone leaves the room and the man is there alone adjusting his cuff links, does the cologne field still exist? As with gravity, the answer is yes. The field still exists; there is just no one to experience it.

## The Physics of Life: Everyone Gets a Dose of Gravity

At the beginning of each physics class, Coach Lucido selected one student to show us the solution for each of our homework problems from the day before. He or she would choose a spot on the blackboards surrounding the room and write out the solution, using equations, arrows, and stick figures. If ever a student didn't know the solution, the rest of the class would pitch in, suggesting numbers, equations, and the names of trade schools willing to take high school dropouts. Only if we all got completely stuck would Coach help us out.

Each day I watched my classmates at the blackboard: Carolyn with her perfect postbraces teeth, Ryan with his Hawaiian family vacation tan, Jill with Jaguar keys in her Chanel purse. They struggled to make sense of orbiting planets, figure skaters, and speeding bobsleds. Sometimes I knew the answers before they did. I was getting the feeling that despite the enormous head start in life they appeared to have, gravity was still pulling on them. Maybe we had started life in different places, but we were subject to the same laws of the universe, and figuring them out on the blackboard and in our lives didn't come easily to any of us.

Whenever one of us included gravity in one of our blackboard calculations, the number was the same: $32.2$ ft/s$^2$. There was gravity, pulling on everything with the same acceleration, giving nothing away. Gravity was the comforting equivalent of Sister Anne in the lunchroom—no cutting in line and no free ice cream.

Way back in freshman year in New Testament class, when Sister Eleanor said, "We all think that our suffering is the greatest," I looked around the room at my classmates. Their blank faces didn't register even the smallest wince or worry line. No matter what Sister Eleanor or Saint Paul said, I was sure I had more on my shoulders than anyone in my class. Not of anyone in an orphanage or a war-torn village, but certainly in that classroom. By senior year, though, I knew that my classmates' lives weren't necessarily simpler than mine. One lost a little brother to leukemia, another's parents divorced in her sophomore year, and yet another classmate's family declared bankruptcy. I was at her sprawling house with a grand piano in the living room and a chandelier in the entryway when their water was turned off because they didn't pay the bill. She didn't look up at me as we silently washed our hands for lunch in her little sister's baby pool in the backyard. She transferred to a public school the next year.

A few girls at school got so skinny that their hair thinned and their cheeks poked out in a scary way. They sipped diet sodas for lunch and gripped their pencils with bony fingers in pursuit of the perfect size and report card.

I began to picture a long algebraic equation that I hoped

would balance out over a lifetime. On the left side of my equation was a finger my dad lost in Vietnam before he knew me; on the right side was the money his Veterans Administration checks earned for college now that I was his daughter. Back on the left side were the mysterious seizures that injured my mother and baffled doctors; on the right side were the many permission slips on which I'd penned her loopy signature with her blessing because she trusted me completely—a trust I had earned by taking care of myself when she couldn't. That long equation with good and bad on both sides might equal out over a lifetime. If it didn't, at least I knew we all had something on both sides of the equation.

Even the luckiest people feel gravity. Sure, the Princess of Sweden had the foresight to choose more beautiful and privileged parents, but you're both here now. Maybe she is a silver bullet shot from a saint's pearl-handled pistol and you are a rock dropped from a working class angel's sweaty hand, but you are accelerating toward the ground at exactly the same rate.

## ((( PHYSICS PRACTICE )))

**1. If you drop a 2-pound water balloon out of a window and it's in the air for 1.5 seconds before hitting the sidewalk, how fast will it be going when it splatters? What about a 14-pound water balloon? Extra credit: About how high is your balloon-throwing window?**

ANSWER: The 2-pound water balloon accelerates at 32.2 ft/s/s. That means it gains 32.2 ft/s every second it's in the air. In its 1.5-second flight, it will speed up to 1.5 × 32.2 = 48.3 ft/s. The 14-pound balloon will accelerate in exactly the same way and also have a velocity of 48.3 ft/s when it splashes down.

**EXTRA CREDIT ANSWER:** Since the balloons start with a velocity of 0 and accelerate steadily to 48.3 ft/s, we can find their average speed by dividing 48.3 ft/s by 2. Multiply the average speed of about 24 ft/s and a flight time of 1.5 seconds to get the height of the window: 36 feet. Prime water balloon throwing elevation.

2. **Galileo got into trouble with the Catholic Church by agreeing with Copernicus that the Earth orbits around the Sun and not vice versa. Since the Church considered our home planet to be the center of God's great and glorious universe, this was offensive.\* Given the size of the Sun (huge) and the size of the Earth (less huge) and the law of universal gravitation, why does it make sense that the Earth rotates around the Sun and not the other way around?**

**ANSWER:** All of the planets are in orbit around the solar system's center of mass. Since the Sun is the biggest blob of matter in our solar system, it is very near the solar system's center of mass. So that's what the planets circle. All masses exert a gravitational pull on each other, but the biggest mass wins this tug-of-war. The smaller mass moves a lot and the bigger mass barely moves. So it is with

---

\* In 1992, Pope Paul John II said (and I'm paraphrasing here): "Um, yeah. We botched that one. Oopsie."

the planets. They respond to the Sun's mass in an obvious way, while the Sun responds to the mass of the planets a tiny bit. When it comes to gravity, size matters. The biggest blob wins.

## ⟪ TRY THIS! ⟫

**Have a friend drive a motorcycle straight off of a dock. Make sure he or she doesn't tip the motorcycle up or down, but drives straight off of the dock horizontally. Drop a pebble off the dock at the same time the motorcycle leaves the dock. Which one hits the water first?**

ANSWER: They will both hit the water at the same time. Call 911. How did you get your friend to do that? Have you considered using your powers of persuasion more productively like maybe encouraging people to donate blood? Or maybe next time you could simply drop a small pebble and a large rock off of the dock and watch them hit the water at the same time.

# 8

# CLEARLY STATE THE PROBLEM
## FORCE AND THE FREE BODY DIAGRAM

---

"What do we need to do first?" Coach asked us. He pointed to a question he'd posed on the board asking how much force was needed to raise a freight elevator filled with clowns. Like a well-trained physics platoon, we answered, "Draw a free body diagram!"

By then we knew that to solve a problem we needed to state it clearly. If we don't understand a situation as it is, we can never work on a solution. Learning how to separate and analyze forces like an engineer are essential.

A free body diagram is a good place to start. It is a simple sketch showing the direction and magnitude of each force pushing or pulling on a body. We can use it to understand and isolate the separate forces acting on an object. In the case of the clowns in the elevator, we can draw a picture with arrows showing the force pushing up and force pushing down on the packed elevator to fully define the problem. Coach liked to throw in bits of unnecessary information so we could practice weeding out what mattered and what didn't. There were 3 circus clowns with red hats and 7 rodeo clowns wearing chaps. All but 2 were Capricorns. Their total weight was 2,000 pounds.*

When drawing the downward force of the weight of the clowns, we could safely ignore their performing specialty, wardrobe, and astrological details. All we needed was their total weight.

In the first stages of engineering a system, a free body diagram is traditionally drawn on a napkin. Even if there is a perfectly good pad of graph paper within reach, engineers like to sketch things on napkins. It makes us feel like we are renegade geniuses. We imagine ourselves to be a part of a great tradition: Galileo secretly scratching diagrams of the orbiting planets on his wall during his house arrest, Da Vinci sketching the first

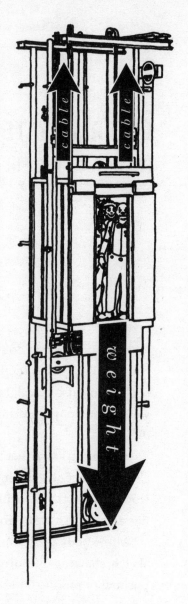

---

* They all drove there together in one small car. Obviously.

design for a helicopter on a wine-stained coaster in Florence, Marie Curie scribbling results of experiments in one of her radioactive notebooks. That first free body diagram doesn't have to be artistic or complex. At this stage, a drawing showing simple stick figures with arrows representing the forces is perfect. With those arrows, we can show the direction and magnitude of forces pushing or pulling on an object. The longer the arrow, the larger the force it represents. The direction of the arrow shows the direction of the force. Easy!

## Statics, the Science of Staying Put

The simplest kind of force problem is one where the forces in each direction are equal and the objects don't move. Structural engineers are experts at designing their beams and struts so that all the force vectors oppose and cancel each other. All the forces pushing on a bridge, or a building, for example, should result in that structure staying in one place.

Let's practice and sketch a free body diagram to show the forces of something staying put. Draw a picture of yourself. A stick figure is fine, but award yourself extra points if you add some party clothes and a few accessories. In the picture you are standing still (with excellent posture). You have no intention of going anywhere. So what are the forces acting on you? Draw arrows for them: one arrow pointing down to show your weight (force from gravity) and one arrow pointing up on the bottom of each foot to show the floor pushing back. Your arrows will be pointing in opposite directions and will cancel each other out. Makes sense, right? The floor will only exert a force equal to your weight. It won't get overzealous about pushing back and throw you into the air.

Now draw an equally stylish friend next to you. You are posing for a picture so she leans against you. Neither of you is moving. So the force of your friend leaning against you is one arrow toward you, and the force of you leaning against her is the same size arrow pushing in exactly the opposite direction against her. The direction and magnitude of the arrows all cancel each other out—you are not moving. You are ready for your glamour shot.

## Force and Acceleration: Twin Action Figures

To understand force and motion, let's return to our rock star Newton, lounging backstage and washing down his apricot biscuits with hard cider. His first law tells us that if all forces are equal on an object or a body, the velocity (speed and direction) of that object or body is constant. That means it's not speeding up or slowing down—not accelerating or decelerating. So in the case of you and your friend leaning against each other, your velocity is at a constant of . . . let's see . . . 0 miles per hour. Yep, that's definitely a constant velocity. All forces are balanced.

Here's another example of all forces being balanced for an acceleration of 0. You are cruising along in your car at a constant velocity of 40 miles/hour. Your car isn't accelerating. That means the forces trying to slow your car down (tire friction, air resistance, the safety cones under your wheels) are equal to the forces working on speeding it up (your engine). If there were no friction, no wind resistance, and no safety cones so rudely placed in your way, you wouldn't need any force from your car's engine to continue cruising along at 40 miles/hour.

When forces *don't* equal each other on that free body diagram, objects accelerate or decelerate. When you sled down a hill, gravity speeds you up until friction or a tree slows you down. Friction will slow you down gradually, with a small force arrow pointing backward. The tree will slow you down suddenly with a big force arrow pointing directly at your open, screaming mouth.

When we get busy assigning numbers to force, we return to Newton's definition. Force requires two things: mass and acceleration. (Remember that acceleration can be negative if the mass is decelerating.) So the equation is simple: $F = m \times a$.

If you know the mass and force, you can find out how quickly something will speed up or slow down. If you know the mass and acceleration, you can find the force. You get the idea.

## Physics of Life: Your Personal Free Body Diagram

The more you know about force and free body diagrams, the more easily you can picture the forces acting on you and your life.

An airplane has four main forces that must be constantly balanced for a successful flight. They are weight, lift, drag, and thrust. These arrows push down, up, backward, and forward. The plane needs the lift to fly, the weight to land, the thrust to move forward, and the drag to slow down and remain stable.

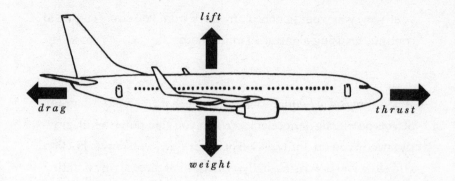

After picturing every action and reaction in terms of vectors and free body diagrams in physics class, I started seeing free body diagrams everywhere. Not just the literal lift, weight, thrust, and drag of airplanes and race cars, but our virtual vectors of lift, weight, thrust, and drag. They are our confidence, fear, ambition, and everyday realities of survival. They lift us up, pull us down, push us forward, and hold us back. For a particular goal, you can diagram the contributing forces on a napkin and examine them. Then, just like an engineer designing an airplane, you can shrink or grow the arrows and apply force where it will be magnified. It seems obvious to do whatever you can to make thrust and lift bigger by killing off fear and doubt, but we need some healthy fear and a good dose of reality for stability. If a pilot didn't intelligently incorporate drag, her plane would do crazy loop-de-loops in the sky before auguring into the ground nose first. If the plane didn't have weight, it would soar into the ether.

In the same way, we need all four arrows to balance in any of our endeavors. We need a bit of restraint and caution. We don't want to head straight into the sun with no way down except to burn our wings off like ridiculous Icarus. On the other hand, if our vectors of fear and doubt are too large, we will sit on the tarmac for years.

Say, for example, you want to start your own dance company to stage inventive modern pieces, but you don't have a fat grant or corporate sponsorship (yet). You will need to balance all four virtual force vectors—up, forward, back, and down. Your upward-pointing arrow is your enthusiasm for the whole undertaking. People with crazier ideas have succeeded; why not yours? You can totally do this! You have a virtual vector of optimism pointing straight up like lift on an airplane. Pulling you down and keeping you cautious is the small voice telling you not to invest every dollar you have in a dance studio just yet. Your ambition pushes you forward. You take action: recruit dancers, create new choreography, schedule practice, and book shows. A healthy dose of reality keeps you from sprinting ahead at a shin-splinting pace. You need to pay your practice studio rent, so you come up with practical ideas for income: ballet classes for kids, disco night for seniors, and pole dancing contests for singles.

Take the squirrely emotion out of it. Just draw the free body diagram and get to work on the vectors. There's something about drawing the diagram that makes our actions less emotional and more doable. You can quickly see if the arrows sum up in a direction you want to move. You are not facing down spooky, supernatural demons. You are simply engineering your life.

# ((( PHYSICS PRACTICE )))

**1. Use a free body diagram to show how your weight is distributed in a flat boot versus a high heel.**

ANSWER: Your free body diagram for a flat boot will show that your weight is distributed evenly—plenty of small arrows pushing up along the length of the sole. Your free body diagram for a high heel will show an arrow pointing up at the front of the shoe (the toe) and the back of the shoe (the pointy heel). Since high heels tend to pitch your center or gravity forward, the arrow up on the toe side will be larger than the arrow up on the heel side unless the heel wearer makes an effort to shift her weight back on her heels. (Or *his* weight on his heels. Anyone can rock high heels with good posture.)

**2. Draw a free body diagram for a particular goal in your life. What are the lift, weight, thrust, and drag? Which arrows need to be adjusted? Sketch your personal free body diagram on a napkin for maximum engineering authenticity.**

ANSWER: Only you know the answer to this one. I would like to run my fastest mile time again. My lift (arrow up) is my freakishly strong desire to do it. My weight (arrow down) is my body, which is now decades older than when I ran that particular mile time. My thrust is my ability to stick to a training program. My drag is all the other projects competing for time and attention and the continuing and annoying need to go to work so I can buy food and pay rent. That drag is a big arrow. I need to reduce it by carving out time specifically for this goal.

3. Calculate exactly what an air bag can do for you using $F = mass \times acceleration$. (The math is really easy.)

You are driving along at 30 miles per hour with your 10-pound head 2 feet from the steering wheel. You pass a person dressed as a hot dog and can't help admiring his or her fantastic sign-spinning skills. As you are wondering if the new hot dog place is as good as their mascot's moves, the car in front of you stops and you rear-end it. Your car comes to a complete and immediate stop.

A. With no air bag, if your head moves straight forward it will hit your car's steering wheel at what speed?

B. If the steering wheel stops your head in a tenth of a second (painfully abruptly), your head hits with what force?

C. How much does it help if it takes half a second for your car to come to a complete stop while its front end deforms and crumples?

D. If an air bag slows down the head's journey to the steering wheel, making it take a full 2 seconds to get there, with what force does the driver's head hit the steering wheel?

ANSWERS:

A. 30 miles per hour. That can't be good.

B. Convert your initial speed of miles/hour to feet/second:

30 miles/hr × 5280 ft/1 mile × 1 hr/3600 seconds =
    44 ft/second
Convert the weight of your head to mass:

10 pounds/32.2 ft/s² = 0.31 slug (standard unit for
mass—ridiculous name but very handy for
calculations like this)

Now find the acceleration of your head:

Slowing from 44 ft/s to 0 ft/s in 0.1 second is a
deceleration of 440 ft/s²

Now you're read to plug everything in to Isaac Newton's
equation:

$F = m \times a$

$F = 0.31 \text{ slug} \times 440 \text{ ft/s}^2$

$F = 136 \text{ pounds}$

OUCH!

C. Find the new acceleration of your head:

Slowing from 44 ft/s to 0 ft/s in 0.5 second is 88 ft/s²
deceleration

Plug the new acceleration into Isaac's equation (the mass of
your head hasn't changed)

$F = m \times a$

$F = 0.31 \text{ slug} \times 88 \text{ ft/s}^2$

$F = 27 \text{ pounds}$

So much better, but still. Ouch.

D. Calculate the new acceleration of your head:

Slowing from 44 ft/s to 0 ft/s in 2 seconds is 22 ft/s$^2$

$F = m \times a$

$F = 0.31$ slug $\times 22$ ft/s$^2$

$F = 6.8$ pounds. Much better. You're tough. You can handle that.

4. "Newtonian" is an adjective used to describe the classical study of mechanics or dynamics. Isaac Newton is honored in this way because he truly invented and organized a particular way of thinking of forces and motion. Used in a sentence, one might say, "The Newtonian worldview is perfectly valid until we study objects moving at the speed of light."

What would your adjective be and what would it describe? Use it in a sentence.

ANSWER: Your answer, of course, will depend on your name and your special skills. My adjective would be "McKinleyesque" and it would describe the great enthusiasm one has at the beginning of a task or project because one has no idea how difficult it will be. Used in a sentence: "After betting several thousand dollars on his own win, he pinned on his racing bib and approached his newly purchased camel with a McKinleyesque confidence even as he wondered how exactly to saddle and mount the magnificent, spitting beast."

# 9

# USE A CROWBAR OR TWO
## MECHANICAL ADVANTAGE

If you are really clever (and you are), you will use what you know about physics to help you with the tough jobs in life—pulling up floorboards, paying for college, and recording your first album.

First, let's make sure we understand how levers, pulleys, and gears work their magic, and then we'll apply that knowledge to our glamorous daily lives. The scientific definition of work is force exerted over a distance. The units for work are foot-pounds. Dragging or pulling weight over a distance is work. Yes, thank you, science. We knew that.

But there is help available. The simplest example of mechanical advantage is the trusty crowbar. If you have done any serious home improvement, you are already great friends with the crowbar. While you exert a modest force on the long side of the crowbar, the crowbar pushes up with a larger force on its short side. Your work (medium force through long distance) is converted to work done pulling up an old wood floor (large force through short distance). The work done on each side is the same, but you placed the larger force where you wanted it—just under those god-awful blue painted floorboards. If you tried to peel the floor up with no help from the crowbar, it would hurt your lovely hands. To calculate the crowbar advantage, you simply multiply force and distance for each side of the crowbar on each side of the equation:

Big force on the work side × small distance on the work side = Small force on the person side × distance on the person side

$$F \times a = f \times b$$

So if you have a crowbar with a 4-inch short side and a 2-foot long side and you lean down with 20 pounds on the long side, how much force are you putting on the short side to pop up those old, ugly floorboards?

FS × 4 inches = 20 pounds × 24 inches
Force on the short side is 120 pounds. Those floors don't
    stand a chance.

Levers, bicycle gears, car jacks, and pulleys all work a bit like crowbars. You do work on one side and they convert it to work on the other side. The trick is to know which end of the crowbar to use and how big the crowbar should be for your particular job. On a bicycle with many gears, we essentially have a combination of crowbars available for our use. When we make it easier to spin our pedals, we don't get much work out of the wheels of our bike.

When we make it harder to spin the pedals, we get a lot of work done on the wheel side of the bike.

If you are riding trails on your mountain bike while the sun is setting and you start to get that "I'm being tracked by a hungry cougar" feeling, you'll want to choose your gears wisely. If you stay in too high a gear, your legs will struggle to turn the pedals and you'll slow down. If you are in too low a gear you'll be spinning your legs like mad without going very fast. Either way you will be in the woods long after the sun goes down with that "I'm now definitely being chased by a cougar that I can hear running up the trail behind me" feeling. You want to use just the right amount of mechanical advantage for each uphill, long straightaway, and downhill to get out of the woods before it's dark.

## The Physics of Life: Crowbars Everywhere

Archimedes famously said, "Give me a large enough lever and a place to stand and I will move the Earth." He didn't realize that there is no place to stand in space and nowhere to rest the fulcrum of a lever out there, but his point is made. With enough mechanical advantage, you can move anything. The trick, though, is to know when to use mechanical advantage and how much to use.

> With enough mechanical advantage, you can move anything.

Since Archimedes' time, javelin coaches and productivity gurus have been instructing us to work "smarter, not harder." The best way to do that is to recognize and use the crowbars in your life.

One day while I was waiting for my mom to pick me up after school, I was talking to Sister Eleanor in her office. I mentioned I was hungry. She took an apple from her desk, held it out to me, and said, "Ask and you shall receive." It was a small act of kindness that reinforced the New Testament lessons from freshman year and reminded me that the people around me can't know what I need if I don't tell them. I had become a stoic little creature during the tough years in Alaska. I ate the apple while I sat on the curb outside and waited for my mom to pull up. I thought, "Well, that was easy."

Years later, when I wanted to record my first CD, I asked experienced musicians for their help. To my surprise, they all agreed to play on my recording. I couldn't pay them much, but I paid what I could and brought homemade lunches and cookies to the studio. The talented musicians showed up, ate peanut butter cookies, and played their hearts out. We instinctively know that we must take turns using a crowbar and being the crowbar. When other young musicians wanted to open shows for me later, I was their crowbar.

There's no point trying to take on the big challenges alone with your bare hands. Find a fulcrum and a lever. Pay in gratitude and cookies if that's all you have. All this hard work doesn't have to be such hard work.

## ((( PHYSICS PRACTICE )))

**1.  If your friend sits on one end of a seesaw and you try to lift her by pressing down with your arms on the seat on the other side, will it be easier or harder to lift her if she moves closer to the middle of the seesaw?**

ANSWER: It will get easier to lift her as she scoots closer to the center of the seesaw. You are doing less work by moving her weight through a shorter distance because she doesn't start as low or end up as high. As long as you press on the very end of your side, you will get the mechanical advantage, aka "crowbar effect," on your side.

**2.  When she gets close enough that you can lift her, how do you calculate how much force you are using on your end?**

ANSWER: When she gets to a position that allows you to lift her, you can calculate the force you used like this:

Remembering the lever equation:

$$F_1 \times D_1 = F_2 \times D_2$$

If you call her side 1 and your side 2, you are looking for $F_2$. Her weight would be $F_1$. The distance from your friend to the center of the seesaw is $D_1$. The distance from where you are pushing down to the center of the seesaw is $D_2$.

$$(F_1 \times D_1) / D_2 = F_2$$

So:

(Friend's weight × distance from center) all divided by your distance from the center will tell you how much force you are using to push down and lift her.

Note: If you ask your friend's weight while you are struggling to lift her while pressing down very hard on your side of the seesaw, she may abruptly jump off and stomp away while you are still pushing down. To prevent injury to your hands and head should this happen, you should wear thick gloves and a helmet. Or you could covertly look at your friend's driver's license and add 20 percent to the weight shown on it instead of rudely asking her while you are red-faced and straining to lift her.

# 10

## LOVE YOUR ROUGH SPOTS
### FRICTION

At the chalkboard, Coach Lucido was describing the force from friction. He sketched cars skidding to a stop, bricks sliding out of the back of trucks onto the freeway, and motorcycles spinning out of control on black ice. Between the mistrust Coach was instilling in us for the fickle grip of friction and the gory snapshots Mrs. Mendez had already shown us in driver's education, we were resigned to live the next few years to the fullest before our inevitable death at a poorly marked four-way stop.

The force from friction was another arrow to add to our handy free body diagrams. It was an arrow pushing backward while an object scraped forward against a surface. A deeper understanding of friction would help us create more accurate free body diagrams. It made sense to us that a rough surface provided more traction than a smooth surface. If you tried to push an empty wooden crate across a concrete floor, for example, you would feel force pushing back from the friction between the crate and the floor. That friction force would increase with a stickier floor or a heavier load in the crate.

If a chubby monkey were to climb in the crate, the weight of the load would increase and the friction between the crate and the floor would be greater. You would have to put more muscle into that crate before it started moving. (Besides adding weight and increasing the friction force, the monkey will likely throw orange peels and whatnot at you while you try to push the box.)

You can make your job easier by slathering the floor with coconut oil. With a layer of oil between the crate and the floor, there is less "rub" between them. Better yet, the monkey will be content to lick the coconut oil off of the floor and stay out of your hair altogether.

This stickiness or roughness between the concrete floor and the wood crate can be pinned down to a number. It is called the coefficient of friction. For wood scraping against concrete without any coconut oil or grease, the magic number is 0.62. When concrete and wood rub on each other, you can multiply the weight on that rubbing surface by 0.62 to find the friction pulling back on the crate as you try to get it to move. That 0.62 is a number with no dimensions that engineers find by experimentation. In fact, Da Vinci was one of the first to find coefficients of friction. (It wasn't enough for him to draw, sculpt, make maps, design helicopters, and sketch human anatomy. He had to be a damn fine materials engineer as well. Apparently he was also very good-looking. We can only hope that his bathroom was a mess.)

There are extensive tables showing the coefficient of friction for aluminum, cast iron, brick, glass, wood, ice, wet snow, dry snow, leather, hemp rope, and every other kind of surface one can imagine. Looking at these tables always makes me wonder what scenario would possibly require us to know such friction relationships. I picture Vikings running out of saunas in their leather underwear and sliding down a snowbank for a dip into the North Sea, then pulling themselves by a hemp rope out of the frigid water before running back into the wood-floored sauna, sliding to a stop just short of a pile of hot coals. Yes. To execute this perfectly it was important for them to know all those coefficients or at least have a pretty good gut feel for them. Thank goodness for that inclusive list of coefficients of friction.

Like those Vikings on a spa day, even without knowing exactly how to calculate it, I'd already put my knowledge of friction and traction to work as a cross-country runner. By the time I was a senior in Coach Lucido's class, I was in my fourth year on the varsity team, but in my freshman year, getting on the varsity cross-country team was a fight. After school, the other girls and I would run the dusty cow trails near our school in a pack. We charged the hills like crazed children's crusaders and sprinted the last stretch of sidewalk to our school. I was the seventh- or eighth-fastest girl, depending on the day. That difference was crucial because only seven girls could be on the varsity team. Cross-country turns the solitary endeavor of distance running into a team sport by its scoring system. At the end of each race, the finish place of the top five runners on each team are added up and the low score wins. Each team is allowed to enter seven runners in a race. The additional two runners are there in case one of the first five trips or gets lost on the course.* The girl who sometimes beat me for that seventh spot was a senior, so it was understood that she should be on the varsity team instead of me. Also, she was part of the Martin Dynasty.

The Martins were a family with four daughters who dominated sports at my school. I was pretty sure they were blood doping at breakfast, but it wasn't my place to say. Carol Martin, the oldest and scariest, had been a star runner when she was younger. She was now a senior and didn't like being passed by a skinny, pigtailed freshman. She made that very clear with her postrace meltdowns.

The Martin girls and the rest of the varsity team had parents who planned their summer training schedules and bought them the very latest high-tech training and racing shoes. By the end of my freshman year, the girls all had fancy, feather-light racing flats. They were almost completely smooth on the bottom with a white top. I wanted a pair, but I couldn't ask my parents for racing shoes when I already had a pair of trainers. My trainers weren't feather-light; they were sturdy. They weren't smooth; they had nubs on the bottom (*so* last season).

I beat Carol Martin in enough races to replace her and join her sister Chris on the varsity team. I lived in fear of Chris elbowing me into a duck pond or tripping me during a track workout. It

---

* It happens. The courses can sometimes be unnecessarily complicated to make them exactly the right length. Instructions from the home course coach often sounded like this: "Go around the lake, under the crossbars, take a U-turn around the walnut tree, then back around the lake, this time clockwise and to the east of the crossbars. *Do not run under the crossbars this time.* When you exit the playground area, look for the trail leading to the restrooms. *Do not take that trail.* Run across the lawn to your left. This part of the course will not be marked with cones or flags. Good luck, ladies!"

was raining on the day of our final race—not a polite autumn sprinkle, but a steady pelting, turning the dirt trails of the course into rivers of mud. There was some talk among the coaches of postponing the race, but it was cross-country. We were supposed to be prepared for run in anything. We lined up, the gun went off, and we splashed through the mud, more than a hundred girls jabbing our pointy elbows into one another. A mile into the race, girls around me were wiping out in their slick-bottom racing shoes, piling up in cursing mounds of muddy limbs. Not just the public school girls were swearing. The girls from my school were working baby Jesus and his virgin mother into their expletives in a shockingly inventive and practiced way.

I started up a long hill and fell. I got up, took a few more steps, and slipped again. This time when I got up, my body remembered my years of practice on icy Alaskan trails and playgrounds. I instinctively sidestepped up the hill and cruised over the next flat stretch in a low, splayed-foot scramble. On the next downhill, I leaned forward and let my legs freewheel as long as I could before I had to crunch down like a skier and slide on one foot and then the other. It was like running through the backyards of my Anchorage neighborhood during the messy melt of winter. I knew how to do this. I had a scar in my chin where the Alaskan ice bit me years ago. I'd learned how to stay on my feet since then. Where I could, I ran slightly off the trail in the sticky ground with drowning blades of grass poking through the mud. My cheap shoes with the front nubs gripped like snow tires. I was passing girls who had crushed me all season as they wiped out in their expensive slick-bottomed racing shoes.

I skidded over the finish line, right behind Chris Martin, helping our team win second place in one of the most competitive regions in the country. Waiting in line for our places to be recorded, Chris reached behind her and grabbed my hand for just a second. Enough to tell me I'd earned my place on the team.

## Physics of Life: Personal Friction Factors

To modify Nietzsche slightly, that which does not kill us gives us traction for the next time. We earn rough scars and calluses from our wipeouts. We need these abrasions on our feet and our hearts to give us the grip we need for our next try. We make mental friction tables. We assign numbers to the Carol Martins of the world, a crazy boss, and a kiss so insincere it freezes our mouth. Then we use our hard-earned treads to squeal around corners faster than we ever could without them. And when even your rough spots aren't enough, when you hit black ice, it will feel familiar because of all your previous spinouts. You know what to do. Don't panic. Pump the brakes. Find your traction. You're not driving off the road, not this time.

### ((( PHYSICS PRACTICE )))

1. A 10-pound steel block sits on a steel table. The coefficient of friction between the two steel surfaces is 0.8.

  A. How much force is required to get the steel block moving forward?

  B. Will the coefficient of friction between the same steel block and steel table be greater or smaller if the surface is covered with olive oil?

C. Why would someone cover a steel table with olive oil?

D. That's weird, right?

ANSWERS:

A. 10 pounds × 0.8 = 8 pounds.

B. The block will slide more easily, so the coefficient of friction will be smaller.

C. Maybe to crush garlic with the steel block on the steel table.

D. It depends on how many people you're feeding. Using a giant steel block to crush garlic on a steel table covered in olive oil might be a genius idea. No judgment here.

2. Which of these combinations would you expect to have the highest and lowest coefficient of friction?

A. Rubber boot soles on ice

B. Rubber boot soles on dry concrete

C. Rubber boot soles on wet concrete

ANSWERS: B would be the highest (least slippery); A would be the lowest (most slippery).

## ⟨ TRY THIS! ⟩

Place two paperback books on a table next to each other, the left book with its spine on the left side, and the book on the right with its spine on the right side. Move them closer together while you flip through the pages of each and shuffle them together like you would a deck of cards. Let a few pages of one book fall, then a few pages of the other until each book is closed and their pages are layered on top of each other. Now try to pull the books apart. Why is it so difficult?

**ANSWER:** When you try to pull the books apart, every page of the first book that is pressed against a page of the second book resists your efforts with the friction of paper on paper. One or two of these pages wouldn't add up to much, but a whole book full of pages multiplies the friction force many times. As you try to pull the books apart, picture all the little arrows pulling against you. The pages have strength in numbers. This clearly illustrates the cumulative force of friction and the power of the written word.

# 11

# CHECK YOUR DIRECTION
## MOTION AND MOMENTUM

It is possible to stay out of jail without understanding the laws of motion and momentum, but if you want to ensure an absolutely squeaky-clean arrest record you should be familiar with them. Let's revisit Newton to find out how to avoid a regrettable mug shot.

By now we have left rock star Isaac alone long enough that he has started a small fire in his backstage dressing room. Soon he'll ask the limo driver where to buy absinthe and start ranting about how Leibniz is a fraud and is totally copying Newton's puffy mullet hairstyle. (Honestly, can't we just say that you both invented calculus at the same time and leave it at that? It's not always about who is the first or the prettiest. Well, most of the time it is. But not this time.)

Let's pacify our pouty Newton by assuring him that we remember his first law: a body in motion will stay in motion; a body at rest will stay at rest. In other words, unless an object feels some kind of pushing, pulling, or knocking-off-course force, it will keep moving in the direction and at the speed it's currently moving. This "body in motion stays in motion" bit is the reason why seat belts are so crucial. When you are driving along at a reasonable 40 miles per hour and need to slam on your brakes to avoid mowing over an adorable skateboarder who can't see through his moppy bangs, your car will feel the force of the brakes and stop. No problem. Your body, however, didn't feel the brakes. It is obediently following Newton's first law and continuing forward because no force was exerted on it at all. Luckily, as your body plows ahead at 40 miles per hour, your seat belt yanks back on your chest, and your face and the windshield are denied the pleasure of getting acquainted.

Anyone who has driven home from the store with grocery-filled paper bags in the back of their car is already an experienced manager of Newton's first law. We know that if we make a hard right turn, the bags will fall over to their left. That's because no force was exerted on the bags to change direction with the car. Those bags of groceries were moving straight ahead, so they continued straight when the car turned beneath them. The bags tipped over and the apples made a run for it.

But now that the bags have already tipped over, you can simply hit the brakes if you need a snack. Those apples will continue moving forward and roll right up to you in the front seat. If the groceries are going to follow Newton to the letter, so can you.

## The Moon's Impressive Momentum

Speaking of Newton and apples, let's circle back to Newton getting bonked on the head with a Granny Smith, which inspired his epiphany that the Earth was pulling on the Moon. A question remains. if the Earth is pulling so hard, why doesn't the Moon come crashing down on us? Newton has an answer: momentum.

The Moon was likely first created from a collision of a hunk of matter with the Earth. At that point, the Moon was a random chunk of debris in motion. It was spinning and moving away from the Earth. It couldn't zip off very far into space because the Earth's gravitational pull wouldn't let it. It remains spinning and remains in motion, trying very hard to take off into space, but it can't go far from Earth. With no force out in space to push or pull on the Moon except for Earth's bossy gravity, the Moon will obey Newton's first law and stay in motion, spinning and flying through space, but it will stay close to Earth while doing both. The Moon is like a tennis ball tethered to the top of a pole, then hit with a tennis racket. Instead of flying off away from the pole, the tennis ball spins around the pole. In the Moon's case, the rope is gravity and the original smack with the tennis racket is that mystery collision a long, long time ago. There is no friction loss in the vacuum of space, so nothing to slow down the Moon. This perfectly balanced tug-of-war keeps the Moon right there where we've come to depend on it for tides and poetic inspiration.

## Momentum Keeps on Keeping On

"Today you will prove conservation of momentum," Coach announced on a sunny afternoon as he wrote a simple equation on the board. "The work of our friend Isaac Newton leads us to this conclusion: total mass times velocity *before* a collision equals total mass times velocity *after* that collision." On the board, he drew a diagram of two billiard balls rolling toward each other (before the collision) and then a picture of them after smacking into each other going in different directions (after the collision). He explained that momentum of an object is its mass times its velocity. He then promised us that if we added up the momentum of both balls before the collision and the momentum of both balls after the collision, we would see that the total was the same before and after. They might be going in different directions now, but their precrash sum was equal to their postcrash sum. Why? Because Newton says so.

Encouraged by Coach to question Newton's authority, we dug through drawers of metal bearings, toy cars, and marbles to find collision-worthy opponents and prove conservation of momentum. We weighed our objects on a postal scale, staged tiny disasters, and then measured their postwreck speed and direction.

The effects of our crashes weren't always easy to measure, but the concept was incredibly simple: the total momentum before the crash was the same right after the crash. Energy can be transferred from one object to another. All of it is transferred except for the energy that is used to make a big dent in your car door, create heat as your tires squeal to a stop, or any other energy-burning activity that might take place during a collision. Since momentum of an object is its mass multiplied by its velocity, one can see how a small object moving very fast can have has as much momentum as a large object moving very slowly. In fact, if a small object has enough speed going into a crash, it can push the large object around quite a bit. Let's be engineers and apply that knowledge to make the world a safer and more stylish place.

## Using Momentum to Fight Crime

If you become a tough inner city police detective with mirrored sunglasses and an undeniable charisma, you will need to get used to people landing on the hood of your moving car. I know this because I watch television shows where this happens quite regularly. When this occurs, quickly review your options with conservation of momentum in mind.

If you see that the face pressed against the outside of your windshield is that of your plucky partner, a crucial informer, or your helpful neighborhood rabbi, you'll want to very slowly apply the brakes, making sure to not stop too quickly. The momentum of your car will change slowly enough to allow the ally to cling to the hood of your car until you come to a gentle, complete stop. They can then get off of your car and wet their pants with quiet dignity rather than soil themselves while screaming on your hood.

If, on the other hand, the person on the hood of your car has a clearly marked villain scar over his eye and is aiming a 9mm at your head, you might consider slamming on the brakes as hard as you can. This will send him flying forward and make aiming for your head much trickier for him than it would if you continued at a steady speed. Another option: consider a quick turn. This will send your pistol-wielding passenger rolling off the hood and keep you moving away from him.

Another situation that you will face if you are Detective Brock Chambers or Agent Tiffany De La Mar is the inevitable fistfight on top of a moving train. This is a scenario that pops up regularly in movies about detectives, so it must be a pretty pervasive eventuality in the law enforcement world. Remember, as the swinging and high kicking begin, you and your opponent are racing along at the same velocity and in the same direction as the train. You can test this by jumping up in the air. You land in the same spot on top of the train. This makes sense. You wouldn't expect to jump up in the air and land in a different spot on the train unless the train slowed down or sped up *while* you were in the air.

To use momentum to your advantage, position yourself so that you are facing the front of the train and your opponent is facing the back of the train. That way you can see which direction the train is turning. When the car you are fighting on is about to make a turn to the left, that would be a good time to execute a high roundhouse kick with your left foot. This will push Igor Eye Scar to your right, just as the train turns to the left. Since his body will already

want to continue straight ahead, if you give it yet another nudge in that direction, he will fall off. Certainly don't kick him with your right foot. That could send you off the train as it is turning to the left.

As you can see, it's not enough to understand criminal law if you want to be a successful detective. You must also understand Newton's laws of motion; also, you should know how to run in a tuxedo and/or high heels after your suspect has fled a coronation ceremony or White House dinner party on foot. It happens. A lot.

## No Means No: A Foot Planted Solidly in the Groin Means No, Thank You

Even if you don't have plans to become a streetwise, fistfighting undercover detective with a questionable past and a heart of gold, you may still need Newton's laws to fight crime. What if, after modeling in a runway show for a new designer's evening-wear, you are followed to your car by one of her assistants? You are dead tired from catwalk struts and quick backstage changes. He asks if you would like to have a drink. You politely decline. He insists. You decline again. He insists that he's not hitting on you.* He says that you are a snob and that you think you are better than he is because you are a model. You think, *No, I'm better than you because I have manners and I'm not pushy and desperate.*

Suddenly he lunges forward. You know what to do. You drop

---

* If someone insists they are not hitting on you, they are totally hitting on you. If it must be said, they are already thinking about it.

to the ground on your side, one foot raised, and before he can stop his forward momentum, he charges groin-first into your heel. He crumbles to the ground, conveniently positioning his head on the pavement near your raised feet. He manages to grab one of your feet and you pound his head with your second foot like a piñata until he lets go of you. Then you stand up and call 911 on your cell phone because that poor man clearly needs medical attention.

If you are tempted to feel bad about his injuries, remember, that first strike wasn't from you. It was all him. You simply placed your foot in the way of his momentum-carrying body. In basketball, that is a charging foul against him. You would get a free throw. In this case, you get to direct the police to a person who is in the fetal position and has just developed a lifelong fear of cropped Italian boots and a new respect for a clearly stated, "no, thank you."

## The Physics of Life: Check Your Direction

Momentum isn't made up only of size and speed. It has direction, too. When Newton said "a body in motion tends to stay in motion," he added "in the same direction."

When I found myself in jail for a short time, I saw examples of what Newton was talking about when he added "in the same direction." I was working for an investigative TV show at the time and we may have overstepped our bounds while trying to get the details of a story. While my employers were busy sorting out the difference between "criminal trespassing" and "sightseeing with a camera crew," I made a few friends in the county jail holding area. They were repeat visitors with helpful tips about posting bail and toilet etiquette. As I got to know them, I quickly discerned a pattern. It wasn't one dumb thing that landed each of my new roommates in jail; it was an impressive list of dumb things. Once they accumulated some momentum toward a life of bouncing into and out of incarceration, it was difficult for them to change directions.

At one point in their lives some of those people in jail thought, "Huh, I wonder what it feels like to smoke meth?" It's a fair question. One does wonder about such matters on occasion. But even if you haven't seen a dentally and dermatologically disastrous Faces of Meth photo series, you can pretty easily guess in what direction your life is headed if you start smoking something made in dirty basements with lighter fluid, melted antihistamines, and a dash of Drano. It's an operation that occasionally requires post-explosion cleanup from people wearing Tyvek suits and ventilators. You can imagine the kind of cleanup your life would require.

So, as I consider my financial, social, and drug-taking choices, I try to look ahead six months. How much would the action contribute to or detract from the glamour and success in my life? Does it lead me to elastic-waist pants, overdraft notices, and/or jail, or does it nudge me toward parasailing in a black string bikini backlit by a golden island sunset? I remember the nice guard in jail handing me a sandwich while shaking her head to communicate *This meal was made by working inmates who may or may not have used their bodily fluids as condiments.* At that moment, I was craving a lovely cheese plate with stuffed olives and maybe a room temperature merlot. I pictured the momentum vectors toward that cheese plate and the merlot. They pointed directly away from the "No Trespassing" sign I'd eagerly run past nine hours earlier.

The steps I'm taking may be small ones, but each is a step in a specific direction. I know that once I start in that direction, only a squealing corner or a collision can change my course. I know it's best to choose wisely in the first place unless I want to graduate to full-time inmate and wear a universally unflattering orange jumper.

This "direction of momentum" consideration becomes even more important when momentum is combined. In Coach's class, we worked through problems with objects that stuck together in a crash. If a car rear-ends a truck and their bumpers twist over each other and connect the two vehicles, they become one object. That means their shared momentum is now their combined weight multiplied by their shared velocity. If they were initially going in the same direction, their combined momentum after the connection is impressive. If they were going in opposite directions before the crash, they will work against each other. I always think of this when I'm combining efforts with other people. The first question I ask is "Are we headed in the same direction?"

If you are joining a band and everyone in it can decide on common goals like touring on the West Coast, selling thirty thousand CDs, and getting a song used in a television show about sexy teenage vampires, your group will be a powerful collective pushing in the same direction. If, on the other hand, the rhythm guitarist wants to only play festivals in Iceland, the bass player wants to sign up for a reality TV competition, and the lead guitarist is just waiting for her career as a pastry chef to take off, you will be like cars that all collided in an intersection while headed in different directions. The result will be a twisted mess of parts at a total standstill. (Tip: if anyone in the band suggests matching haircuts, get out immediately. Some of us learned that one the hard way.)

It's worth asking yourself, With whom do I want to team up? Are they really heading in the same direction as I am? If no one shares your vision, don't let that stop you. You can take off on your own. Remember, momentum is made up of mass and velocity. You may be one tiny person, but with a big running start, you can be the bullet that pushes a bowling ball.

### ((( PHYSICS PRACTICE )))

1. **Two figure skaters are warming up on the ice. Glenda is tall and muscular (a good jumper) and weighs 150 pounds. Kronda is willowy and tiny (high points for artistry) and weighs 100 pounds. They are each skating at the same speed. While Glenda is skating north and Kronda is skating south, they smack into each other and slide in a combined ball of stretch fabric and glitter across the ice. If no energy is lost in their collision or in friction on the ice (not possible, but let's pretend for fun), in what direction will they be going after their collision?**

ANSWER: We know that the combined momentum before the crash is the same as the combined momentum after the crash. We also know that the skaters were going at the same speed when they lutzed into each other and Glenda has more mass. (All muscle, Glenda. You look fantastic.) So there was more momentum headed north than there was headed south before the collision. Since the combined momentum after the crash has to be the same, there will still be more momentum headed north when they are a ball of bruises and blades. Thus they are sliding north.

**2. Who wants extra credit? You do!! If Glenda was headed north at 15 feet per second, how fast would little Kronda need to be going to bring Glenda and herself to a dead stop when they crashed?**

ANSWER: Ready for some algebra? Of course you are. You eat algebra for breakfast.

> Glenda's initial momentum + Kronda's initial momentum =
>     zero final momentum
> (Glenda's mass × Glenda's velocity) + (Kronda's mass ×
>     Kronda's velocity) = 0
> Super-helpful hint: Kronda's velocity is negative since she's
>     going in the opposite direction with respect to Glenda.
>     You're welcome.

We'll be mathematically lazy and use weight for mass since it won't make any difference in this case.

> (150 pounds × 15 ft/second) + (100 pounds × (-V ft/s)) = 0

Solve for V and get 22.5 feet/second. See? Even if you are little you can pack a mighty punch and stop a bigger mass if you are moving fast enough. The crash still hurts like hell, but it's nice to be noticed instead of plowed over.

## ((( TRY THIS! )))

Rent a Roller Derby rink. Recruit a few friends. Put on skates, knee-pads, elbow pads, mouth guards, and helmets. Choose skater names such as "Lady Shatterly" and "Johnny Tsunami." Predict the outcome of various collisions with different starting velocities, skaters' masses, and angles of collision. Award prizes for the highest number of correct predictions and biggest bruises at the end of the skating session.

# 12

## LET THE UNIVERSE MAKE THE RULES

During summer breaks in high school, I learned not to ask my parents for money to take the train into San Francisco, go shopping, or meet my friends for ice cream. Mom and Chuck would only look at each other and say something like "Well, that sounds fun" and then return to cleaning the garage without reaching for their purse or wallet. Shopping and day trips to the city were not in their budget. More importantly, they knew that my craving for uniform code–breaking shirts and mint chip ice cream would be the perfect motivator for me to earn my own money. Working during semester and summer breaks showed me exactly what my career options were without a good education. I squirted mustard on endless rows of buns and did yard work for my neighbor. She was an older woman with the charm of a plantation overseer. She owned several grimy little rental properties, so she had plenty of planting and weeding for me to do.

I would return to school in September or after spring break smelling like ketchup and bark chips, motivated to get good grades and get into college. Burgermaking and raking were perfectly acceptable jobs for me in high school, but not something I wanted to do for the rest of my life . . . or ever again. I had glimpsed my future without a degree. It included a paper hat and a slumlord who swore viciously at her oleanders for insufficient flowering.

Like I did during those summer breaks, let's remember our original motivation and revisit why we are journeying into the heart of physics. Let's remind ourselves what happens when instead of following the laws of gravity, energy, and motion laid out before us, we ignore them and make up our own laws. Cult leaders are a fascinating example of this. Not the cons, the ones who intentionally trick people into believing nonsense; they know the difference between reality and what they are selling. It's the other kind of cult leader who is so interesting to watch, the ones who truly want answers and create them from their imaginations.

After announcing that the world will end at midnight on a certain day, the charismatic leader of devoted followers lays out all his supporting evidence. He makes a case using Viking runes, Polynesian statues, Euclidian geometry, and a detailed analysis of his own dreams of flying over Manhattan. He announces to his followers that on the Last Day, everyone must form a single line and walk around the perimeter of the compound making a joyous noise by banging wooden spoons on baking pans. Most importantly, they must truly believe in him, their divine leader. Also, they must wear light blue robes so that the interplanetary messenger gods will recognize them as faithful and deserving.

This part about the blue robes cannot be stressed enough. If you are in a robe of any other color, you will be left behind on a crumbling and incinerating planet, roasting and wailing with the unbelieving and undeserving in robes of pink and paisley.

Midnight approaches, pans are banged with wooden spoons, the perimeter is walked in light blue robes, and the world does not end. Maybe the end of the world is a tiny bit late, the loyal followers think, so they keep up their joyful noise, but their arms are getting tired. By the time the sun comes up, the true believers shuffle back to their quarters, their robes heavy with rain, clutching their dented pans, and wondering if their leader missed something in his calculations.

What happens next is the most fascinating part. The leader announces that he has discovered the error. The world will definitely end two years later on the same day. He forgot to factor in the time change between Norway and Polynesia. The enlightened leader won't admit that the world is not interested in his weird little calculations and will end when it damn well pleases.

Of course, we think these people in their mud-fringed robes are ridiculous and their leader is a freak and a fraud. The truth is, though, even those of us who are not in doomsday cults occasionally insist that the universe follow our made-up rules. We tell ourselves that if we were 15 pounds lighter we would have a better job. We plan outdoor weddings thinking that the rain wouldn't dare impose on true love. We are just like that cult leader who sits in his bunker at night making planetary distance calculations on a legal pad and communicating with his Viking ancestors using a system of smoke rings and eight-sided dice.

The truth is, we can't wrestle the world to the ground and force it to adopt our rules of engagement. We are free to make up any laws we want, but the universe is under no obligation to follow them. Our best chance of steering our lives in the direction of giddy freedom and giggly happiness is to understand the operating principles of this universe and use them to our advantage. Those laws are spelled out when we see how gravity, motion, energy, and matter behave. When we are on friendly terms with them, not only can we escape from sinking cars and execute a gorgeous stage dive, we also can use the laws of physics to create useful models to keep us optimistic, balanced, and looking straight into our glamorous future.

Now let's get back to class.

# 13

# PREPARE TO FLOAT
## BUOYANCY

---

Coach Lucido had my full attention the day he covered buoyancy. He explained that buoyancy is the upward force that makes things float. As fluid is pushed aside by an object, the fluid pushes back to try to reclaim the space from which it was so rudely evicted. All this pushing back makes the object float. That wasn't the part that fascinated me, though. It was the naked guy part.

## The Naked Guy Part

Apparently, the philosopher, scientist, and athlete Archimedes* suddenly understood buoyancy during bathtime. Coach only de-

---

* Rhymes with "Mark My Wheaties," which is a thing nobody ever says.

scribed Archimedes as a philosopher and scientist, but since Greeks liked to be well rounded, it was not such a stretch to assume that Archimedes was also an accomplished discus thrower or wrestler, and if I was going to picture a man in a bathtub, he was going to be ripped.

Archimedes had a flash of insight as he lowered himself into his tub, his rock hard triceps shaking and his abs crunching to support his taut and muscled lower body. Archimedes' soulful brown eyes watched the water overflow from the tub onto the floor. He realized that the water flowing onto the floor was exactly equal to the volume of his chiseled body. As a delicate line of perspiration and steam collected on his whiskered upper lip, he wondered if the overflowing water was somehow related to

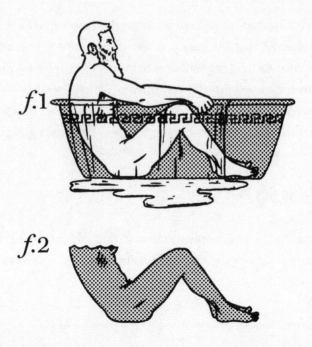

*f.*1

*f.*2

how light his powerful physique (figure 1) felt in the tub. As he scrubbed in slow circles on his broad chest, it came to him: that displaced water (figure 2) was kicked out of its spot and was pushing to get back into the space now occupied by his body. The only force available to the water is its own weight. Thus the weight of the displaced water was pushing on him and lifting him up.

"Eureka!" yelled Coach Lucido. In forceful Kabuki-like squats he pantomimed the buoyant upward thrust of the water under Archimedes. He explained that when water is pushed out of a space, it pushes back with all of its approximately 62 pounds per cubic feet. When you displace 1 cubic foot of water, what will the buoyancy force from the displaced water be? That's right, 62 pounds. If that cubic foot object pushing the water aside weighs less than 62 pounds, the object will float. If the cubic foot object weighs more than 62 pounds, it will sink. The question of sinking or floating is simply a pushing match between buoyancy (the push up from the water) vs. the actual weight of the object bobbing on the surface or sinking to the bottom.

Coach added a twist by telling us that if you displace a fluid other than pure water, the push up from buoyancy will be different. For example, ocean water is about 64 pounds per cubic foot, because the bits of salt in it are heavier than water. If Archimedes decided to pamper himself and add some bath salts, the upward push of buoyancy on his body would be slightly greater than it was in the unsalted bathwater. He would be displacing the same volume of water with his lovely body, but since water pushes back with its own weight and that same volume would now weigh more, the push up from buoyancy is greater. Sure, salt was expensive in those days, but it was worth it. This extra buoyancy was exactly what Archimedes needed after a hard day of studying circles and power lifting.

I wanted Coach to get back to Archimedes so that he might describe the bath scenario in more detail, but he had moved on to various examples of sinking or floating. While he did that, I flipped through the pages of our book and found a disappointing black-and-white drawing of Archimedes. He was solemn, bearded, and, Jesus . . . he looked like Jesus. Actually, he looked like an older Jesus, like Jesus's dad. Oh, God. That would be God. I'd been picturing God naked.

By the time I had quietly delivered a prayer to God apologizing for mentally undressing him, Coach had sketched a short history of shipbuilding on the chalkboard. It went something like this: Our fishing ancestors first made boats out of wood because they saw that wood floated. It was only logical to hollow out a log and jump in. They didn't know that wood was less dense than water and was pushing the heavier water aside to float. They just knew it worked. Also, they knew they caught more delicious fish when they rowed into deeper waters rather than fishing from shore.

The leap from wood ships to steel ships took many years. Long after our ancestors knew how to make steel knives for chopping onions and beheading invaders, they still didn't consider it a good shipbuilding material. Although it was strong, steel didn't strike early shipbuilders as very "floaty." They knew that their fancy new steel-bladed fishing knife would sink straight to the bottom of the sea if it was accidentally dropped during a rushed halibut-gutting, so it didn't occur to them to make a boat out of steel.

It wasn't until they really worked the upward force of buoyancy by making steel hulls big and hollow that they found steel ships could float rather nicely, as long as they displaced enough water. A hollow shell of steel pushed aside the same volume of water as a solid hunk of steel, but it wasn't nearly as heavy as the solid hunk. The upward push from the water was enough to make steel float as long as it was a hollow shell.

Since steel is so much stronger than wood, those buoyancy-savvy shipbuilders were able to make gigantic, indestructible party boats such as the *Titanic*. Well, maybe not indestructible, but able to comfortably seat a full symphony and host a banquet with a roast beef carving station. If a pointy iceberg punctured the hull allowing all that displaced water to gush back in, the buoyancy was quickly lost. With that water inside the hull of the ship, it was no longer pushing up and providing the buoyancy needed for floating. Still, one could enjoy Mozart and a delicious French dip as one sunk into the icy waters. Honestly, haven't we all attended less enjoyable parties?

Wait, what was that deadly iceberg doing there anyway? It, too, was enjoying the laws of buoyancy. Though it's standard protocol for a liquid to shrink and sink when it turns into a solid, that isn't true for water. Water is special.

## Water: Nature's Special Snowflake

When a normal liquid like hot, runny lead gets colder, its atoms slow down and clump together into a dense, solid blob. That dense blob is heavier than liquid lead. We can prove this by dropping a muffin-sized hunk of solid lead into a vat of hot liquid lead and watching it sink. If you drop an ice cube into a glass of water,

however, it floats. This is because water becomes less dense
when it freezes rather than more
dense. This quirk tells us a lot
about the water molecule structure and community.

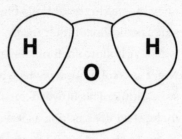

With two little hydrogen at-
oms and one big oxygen atom, the water molecule is arranged
like a V, with the big oxygen atom at the bottom and the two
smaller hydrogen atoms on the other side. When the temperature
drops, water molecules respond with the discipline and order of
British civilians in wartime. While lesser substances cluster together in a random, messy panic when they freeze, water molecules keep their wits. They arrange themselves in a very orderly
fashion. The positively charged side of the water molecule where
the hydrogen atoms hang out steps in line next to the negatively
charged side on the oxygen side of the neighboring water molecule. In this way, they form a lattice structure while maintaining
the proper amount of space between them. (Freezing water molecules, like the British, are not huggers.)

This organized structure of frozen water molecules is more
spacious than its liquid form. A cubic foot of ice contains fewer
molecules than a cubic foot of water; ice is less dense than water.
That means a cubic foot of ice actually weighs less than that same
volume of water. So a block of ice gets a buoyant push up from
water that is greater than its own weight and the ice floats. Since
ice is not much less dense than water, it has to displace a lot of
water to float. It doesn't bounce around on top of the water like a
hollowed-out log. That's why the deadly iceberg that struck the
*Titanic* didn't look so big from the surface. Most of it was sub-

merged to get the buoyancy it
needed to float.

## Using Buoyancy in Your Glamorous Life

All this information Coach
gave us about buoyancy could
come in very handy if you join
an undercover operation to recover priceless marble sculptures
and find yourself on a yacht needing to get them to shore. (It
could also come in handy if you need to get some peanut butter
sandwiches to your nephews on a rafting trip, but let's go with
the undercover art recovery operation because your life is very
glamorous.)

So far in your art recovery operation, you have done an expert
job of infiltrating a group of Russian mafia bosses, Portuguese
pirates, and Canadian dairy farmers all known to traffic in stolen
art. After months of drinking vodka in Odessan nightclubs,
cheering at Portuguese *futebol* matches, and losing money in
rigged church bingo games in Manitoba, you are invited to a
small get-together on one of the suspect's yachts. You depart
from a Spanish port in the afternoon, drink rum from poached
elephant tusk goblets, and play high-stakes blackjack and cut-
throat Pictionary while you sail north.

At dusk, the steward escorts you into a side room so you can
change out of your swimsuit and into dinner clothes. As he is
showing you where the towels are, you see in the half-light the
marble head of the goddess Hera. Next to Hera's head is the foot

of Hercules. You think "Aha! These are the pieces stolen from the Greek Museum last year!" But you don't say that out loud. That would be dumb. Instead, you pretend to be a little drunk and tell the steward that it may take a minute for you to get dressed. Oh, but you're not drunk. You are figuring out how to get those priceless pieces safely to shore.

From the room where you are supposed to be changing, you hear the confirmed art thief speaking to a woman with very impressive cleavage. His plan is to sell the pieces when the yacht gets to the next port in one hour and that tiramisu will be served shortly after that.

As you consider how to get an early serving of the tiramisu and the treasures off the yacht in less than an hour, you realize that the whole party was a cover. That's why you didn't see any lifeboats or personal flotation devices. Mr. Make Your Guest Wait For-freaking-ever for tiramisu didn't want anyone running off with the sculpture pieces in a lifeboat. He probably has a life vest stashed away for himself somewhere. He has no respect for the safety of others, and he stinks at Pictionary. This guy is really starting to get on your nerves.

You quickly assess the situation. You are 2 miles from shore. You know you can swim that distance, but you can't leave without the marble pieces, which are each about 50 pounds. They would weigh you down and sink you like, well, stone . . . because they are stone.

The host already told you and the rest of his guests to help yourselves to whatever you might need in the galley (kitchen, whatever), and so you do. While he and several men with Siberian tiger tooth cuff links are arguing about the Portuguese rules for Russian roulette on the starboard deck, you pull the garbage bags from the galley bins. They are the thick, industrial-strength bags you were hoping to find. You take hair ties from the well-appointed ladies' room. You drop all this off on the port deck. Next, you tiptoe back to retrieve Hera's head and Hercules's foot. With everything collected, you put each of the pieces of sculpture in its own bag, blow the bags up, and wrap the hair ties around the top of each bag to close them.

You then slip over the side of the yacht with the booty in each hand and swim a lovely splash-free sidestroke while pulling your inflated garbage bag away from the yacht. You do all of this in the cover of dark, silently thanking the garbage bag manufacturer for making their bag dead-of-the-night black instead of look-at-me-I'm-totally-stealing-back-the-stuff-you-stole white.

Since you blew the bags up big enough to push about a cubic foot of water aside and water weighs more than 60 pounds a cubic foot, the recovered treasures float easily, with 60 pounds of force holding them up. Pulling them is not hard because there is nothing to stop them once they get moving (as you know from the chapter on momentum). You are a silent tugboat pulling a precious barge while wearing a tiny tiramisu mustache. Eureka and adios!

## The Physics of Life: Preparing to Float

There are times, like when you are recovering stolen sculpture from a yacht, when it helps to understand buoyancy. And there are thousands of other small, less dramatic moments when we must sink or swim. I prepare for them like an iceberg. With only a small part of me visible to the rest of the world, I build an organized structure underneath the waterline. All the sit-ups, homework, research, organization, and vegetable eating I do to build my underwater structure are done in private and without any excitement.

I get very focused on what helps me float and what doesn't. I know that no amount of complaining about how busy I am or wondering what my life would be like if I were born wealthy is going to keep me afloat. The laws of buoyancy don't care how much I think I am suffering. They just care about that crystal lattice under the surface and how much water it pushes aside. All that quiet and invisible effort will create a beautifully organized and carefully constructed flotation device to help me sail to my destination.

If I'm tempted to, just this one time, put off the needed preparation for a test, a gig, or a report on the code requirements for a foam fire suppression system, all I have to do is remember how lack of preparation has worked in the past. Without the required studying, practice, or careful selection of outfit, I have done my share of sinking.

When I see someone interviewed after receiving their first Pulitzer, movie role, or pumpkin-growing ribbon, I love hearing them talk about what an honor it is to be in their position and how all the other short stories, actors, and pumpkins were all so good and how surprised they are to come out on top. I enjoy listening to what they don't say. They don't mention the hundreds of previous drafts they wrote, awkward auditions they bombed, and late night emergency frost tent constructions they navigated on their way to greatness. Now that they are floating, they have that serene look of Archimedes in his tub—a combination of "eureka" and the sweet relaxation of a warm bath. They are weightless with buoyancy.

We know how they got there. They built a huge, invisible structure under the waterline. It's always bigger than we think it needs to be. More practice, workouts, and studying are required

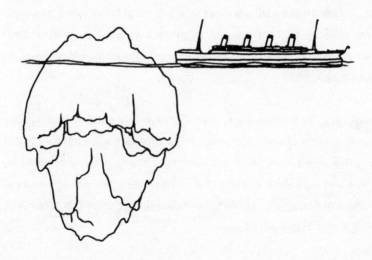

than we thought. I'm still surprised that everything takes longer than I expect and is harder than I imagined. I get tired, frustrated, and wonder if it's this difficult for everyone. Then I look at a photo of an iceberg to remember what it takes to float.

Most of the ice is under the surface, doing the hard work of pushing away the water while the water pushes back and keeps the iceberg afloat. Such a small part of the ice peeks above the water and enjoys the beautiful view.

## ((( PHYSICS PRACTICE )))

1. In one version of the Archimedes bathtub story, he is so excited about the idea of buoyancy and how he can use it to determine the densities of solids, he runs out into the street naked. This isn't really a question. I just wanted you to know that. Discuss.

2. If Archimedes had a brother who had roughly the same measurements as he did, but was not as muscular, would the brother float more easily or sink more easily in the bathtub than the muscular Archimedes?

ANSWER: The brother would float more easily. He would displace the same volume of water, but since fat isn't as dense as muscle, Huskymedes would weigh less and float more easily. (Remember, he has the same measurements, so he has the same volume.) The same force from buoyancy pushing up would be competing against less weight pushing down.

3. There is a well-known but rarely discussed phenomenon that may have contributed to Archimedes's bathtub epiphany. While his arms and legs sunk in the bathtub, the appendage just below his belly floated and pointed straight up. Was this because he was excited about his discovery of the buoyancy principles, or is something else going on here?

ANSWER: The midsection appendage in question, despite some of its more popular nicknames, doesn't have a bone in it like arms and legs do. It is made of spongy tissue (not a very sexy description, sorry), so it's light compared to water. The appendage displaces an amount of water heavier than its weight, and just like your kitchen sponge floats in the sink, Little Archimedes floats in the tub.

## ((( TRY THIS! )))

1. In the deep end of a swimming pool, try to lie on your back and float. After that, fly to Tel Aviv. From Tel Aviv, take a train to Jerusalem. Book a nice hotel. Enjoy room service blintzes. Take an hour bus tour to the Dead Sea from Jerusalem. Apply sunscreen before exiting the bus. Swim out far enough that you can lean back in the water. In which body of water could you float more easily?

ANSWER: You should be able to float easily in the Dead Sea and likely not at all in the clear water. This is because that water your body is pushing aside in the Dead Sea is very salty and heavy. Remember that Archimedes (the totally naked guy who looked freakishly like God) discovered that buoyancy is equal to the weight of the water pushed aside. Heavier water means a more forceful push up.

**2.** In your kitchen, get a can of garbanzo beans or tomato sauce (or whatever you have that weighs about 1 pound) and a large plastic bag—bigger than a sandwich bag. Fill the kitchen sink with water. Now put the can in the bag and blow into it. When it's fully inflated, tie or seal the top. If you put the whole can and bag assembly in the full kitchen sink, will it float? Make a prediction.

ANSWER: Since your can and bag weigh about 1 pound, the buoyancy force pushing up from the water will need to equal about 1 pound for it to float. To do that, you'll need to displace about 1 pound of water. Since we know that water weighs about 62 pounds a cubic foot, you can convert units (with a bit of clever math) and find that a pound of water takes about 27 square inches, which happens to be equal to about 2 cups of space. Does your bag take that much space? You don't have to get too scientific about it; just mentally picture how many cups of air you can fit in the bag. If you can fit 2 cups in your bag, the can and bag will float. Drop the bag and can in the sink. Were you right? Of course you were.

# 14

# LET IT FLOW
## FLUIDS

---

"It's incredibly painful!" Coach Lucido told us cheerfully while explaining the phenomenon "the bends." He described divers ascending from the high pressure of deep water too quickly as the nitrogen in their blood expanded into bubbles. "It can be fatal," he added, still smiling.

Coach was clearly not a good candidate for delivering the news of death to loved ones. If the unfortunate event involved any principles of physics, he was too likely to stand on their doorstep and explain with great excitement how the bridge collapsed from sympathetic resonance, or the fire started because of an electrical short that created heat through a spot of high resistance. Nobody wants to hear that kind of detail in her time of grief, especially not from someone with a gigantic grin on his face.

Coach showed us how the divers in his scenario felt more

pressure the deeper they ventured below the water's surface. It was easy enough to calculate that pressure: the density of the water multiplied by the depth of the diver below the water's surface (plus the atmospheric pressure of the air on top of the water if you want the total pressure). It didn't matter if the diver was 20 feet deep in a small swimming hole in New Mexico or 20 feet deep in Lake Michigan, that diver would feel the same pressure.

Coach Lucido then helped us understand water pressure with a practical example, another in the Let's Terrify Young Drivers series. While he sketched on the board and described how to handle your car sinking in a river, I worked on applying the information to my future life. I'm pretty sure this is not how Coach delivered the lesson, but this is what I heard.

## Keeping Your Head and Shoes in a Sinking Car

If you are driving to a dinner party one evening, swerve to avoid a child playing in the road, drive into a lake, and find that your windows won't open, you'll need to remember what you know about water pressure to make a safe and stylish escape.

If you didn't understand the physics of fluids, you would be in danger of making a real mess of things. Your initial swearing and door kicking would make the veins in your neck stand out in an unattractive way and ruin your shoes. When you were finally able to open the door, you would frantically swim to the surface. Exhausted and disoriented by your earlier freak-out, you wouldn't know which way to swim. One shoe would fall off while you fought to stay afloat. Next, rescuers would arrive and drag you onto a small, inflatable orange boat while your dress hiked up or your pants fell down to expose your drooping underwear. Finally, slack-jawed and shivering, you would grip the handles of the ridiculous kiddie boat as it nears the TV cameras onshore and your ex recognizes you on the live local news.

But thank goodness you do have an understanding of the physics of water pressure surrounding a sinking car so you can handle this unfortunate situation with style. After you try the window and it doesn't work, you'll quickly determine that you must move on to Plan B. You see that the water on the outside of your car desperately wants to get inside of it and fill all that waterless space. Since the water doesn't have hands to open the door, she stubbornly smashes herself against it, trying to break her way in.

Since you know that the water is pushing with all of its weight on that door and you know that water is surprisingly heavy, you quickly determine that despite your ability to do some impressive yoga, you can't push hard enough to overcome the pressure it is exerting on the outside of the car door. You recognize that when enough water has seeped into the car through various non-watertight openings and the water level inside the car rises, the water pressure inside and outside the car will match and you'll be able to swing the door open.

You know you have some time to kill before enough water enters the car to balance the pressure outside the car, so you take this time to prepare for your escape. You remove your shoes and tie them to your belt by their laces or their slender ankle straps. When the water reaches your chin, you take a deep breath and swing the car door open. With water on the inside and outside of the car, the pressures are the same on either side of the door and there is little resistance. You exit the car with a graceful, powerful dolphin kick. When you reach the surface, you use long, slow strokes to keep yourself afloat and propel your body toward the shore where, by now, enough firemen to fill a fund-raising calendar are watching you in admiration. You acted so quickly and with such finesse, they didn't even have time to get the little orange rescue boat out.

You emerge from the water and slip your shoes back on (because they really pull the whole outfit together). Then you smooth your hair back in place while a fireman wraps a blanket around you. The television crew arrives, sets up flattering lighting, and streams live footage of you answering questions about your escape with a caption that reads, "Courageous driver with stylish shoes cheats death and spares life of a child." The mother of the child embraces you. The reporter asks you questions and you explain to him that no, you don't consider yourself a hero . . . unless a hero is someone who selflessly puts themselves in danger to protect a young life and then demonstrates the power of scientific knowledge and a cool head in an emergency, then yes, I suppose the word "hero" would objectively apply in this situation.

## So Much More Than Water

We normally think of fluids as liquids—something we can drink, swim in, or use to lubricate an engine. But to scientists and engineers, the word applies to any continuous substance that can flow. A fluid doesn't crack or break when we press on it, but shifts into a new shape without complaint. By that definition, liquids and gases are both fluids. We describe the resistance of a fluid against deformation with the word "viscosity." The higher viscosity a fluid has, the more it fights back against deformation. Water will very easily take a new shape if you push on it, so it has a low viscosity. Motor oil is a little harder to push around, so it has a higher viscosity than water. Cold motor oil has an even higher viscosity.

Motor oil is one of the normal fluids that earns the right to be called a Newtonian fluid, after Isaac Newton, of course. (He had his hands in everything.) Honey is a normal Newtonian fluid. If you spread honey with a knife, its viscosity doesn't change because you are applying force. The layers of honey flow past each other in an unchanging way no matter how intense you get with that knife. As long as the honey stays at the same temperature, it resists force in a linear way. You push on it and it responds. You push on it more and it responds in the same way it did before.

The weird fluids, the ones that don't behave in predictable ways, are stuck with the label "non-Newtonian." This means they don't flow in the same linear way as those normal Newtonian fluids. Non-Newtonian fluids such as quicksand and cornstarch slurry get less flowy if you hit them with some serious force. In fact, when you smack them, the friction between their layers of fluid increases so much that the whole blob acts a bit like a solid. You can run across a vat of cornstarch slurry, but if you stop, you'll sink. So strange. So very non-Newtonian. With ketchup, another Non-Newtonian fluid, the opposite happens. When we apply force (banging on the bottom of that bottle) the layers get less clingy and the blob of ketchup gets less viscous, acts more watery, and comes flying out of the bottle. It's the same with toothpaste—a little push and it gets all gooey and submissive. It goes without saying that proper Newtonian fluids don't let their kids play with the unpredictable children of non-Newtonian fluids.

Non-Newtonian fluids have every reason to be proud, though. Inventors are looking for ways to use them for nonrestrictive bullet-stopping clothing, and chefs rely on plenty of non-Newtonian fluids to make many essential desserts such as crème brûlée and chocolate mousse. (Yes, they are essential. That is not a question. That is a statement of fact.)

## WWBD: What Would Bernoulli Do?

Let's focus on the Newtonian fluids for now, the everyday stuff—air and water. It is a strange, counterintuitive fact that a faster-moving fluid has lower pressure than a slower-moving fluid. Coach Lucido had us prove the concept by blowing over the top of a piece of paper while holding it lightly on each side. The paper would rise up, drawn into the area of faster-moving air. To drive the point home, Coach fired up a blow-dryer, pointed it straight up, and dropped a Ping-Pong ball into the fast-moving stream of air. The little ball bounced around magically in the stream of moving air, tapping against the high-pressure walls of still air on each side of an invisible column.

When the class dug into Bernoulli's equation—the answer to all questions involving air,* water, and other fluids—it all made sense. I'll spare you a full dissection of Bernoulli's principle, but here's the quick summary: Fluids can have three kinds of energy—

---

* As long as the air isn't moving faster than about 330 feet per second; then there are more involved versions of Bernoulli's principle that apply.

speed, pressure, and/or elevation, and they can be converted from one to the other. Think of water piped from a mountain lake into a city below. The water begins high above the city. It has elevation and the energy associated with it (like a diver poised on a platform). As the water flows into the pipe and downhill, that elevation energy is converted into speed. When the water fills the water distribution piping in the city, it is contained and some pressure builds up. When you turn on your faucet, out comes the water with a bit of pressure. If you wanted to convert some of that pressure back into elevation, all you'd have to do is turn on your sprinkler. Pressurized water pushes up and sprays your yard.

If you're thinking this all has a familiar ring because it sounds a lot like the chapter about potential and kinetic energy, then you need to enroll in engineering classes. Seriously. We need more engineers, and you are connecting the dots beautifully. If this doesn't sound at all familiar, that's okay. I really like that shirt you're wearing. It's a good color for your eyes. Everyone is appreciated here.

Another way to see Bernoulli's principle at work is to choose a window seat on a plane where you can see the wing. If you look at the shape of the wing, you can see how the air that flows over the top of the wing has to move faster to make its way over the curved top. Faster air on top of the wing also means lower pressure on top of the wing. And what does that mean? Higher pressure under the wing and liftoff!

## The Physics of Life: Here Come the Helpers and the Haters

You will clearly see the relationship between velocity and pressure whenever you start a new project. Whether it's a trip to Poland or making a film, once you have made your initial plans and picked up some speed, your virtual pocket of high velocity and low pressure will draw in all kinds of criticism, imitators, offers of help—all things that would never have appeared if you weren't moving at a considerable clip. Be prepared to ignore the haters and downers and welcome the helpers and cheerleaders.

If you are waiting for funding or feedback, I'm here to tell you that you won't get it until you are in motion. The rush of your ideas and forward motion pulls in what you need, but you have to start on your own. And once you are moving ahead, someone will inevitably say, "Oh, it's impossible to find a decent camera operator," or "Poland? I hear the cabdrivers are all opium dealers." That's a good sign. That means you are headed toward your goal and that you are moving fast enough to pull in both criticism and help. The endorsements, funding, and whatever else you need to propel you forward are on their way.

Nice job. Keep moving.

## ((( PHYSICS PRACTICE )))

**1. Where will you feel more pressure, 10 feet under the surface of a 4-foot-diameter seawater tank, or 10 feet under water in the Atlantic Ocean just off of La Palma in the Canary Islands?**

ANSWER: If the water density is the same, you will feel the same pressure. In each situation you have 64 pounds/ft³ × 10 feet = 640 lb/ft² of pressure on you. If the tank had freshwater in it, the density of the water would be different and you'd feel more pressure in the ocean. If the temperatures were different enough to affect the water density, that would also change the answer. But the pressure you feel is only dependent on the density of the water (how heavy it is) and your depth. The surrounding water doesn't make a difference except that you are likely to feel crazily claustrophobic in that skinny tank and much more relaxed swimming off of the Canary Islands.

2.  What's more viscous at room temperature?

    A. Black coffee or cream?

    B. Brandy or vodka?

    C. Gasoline or motor oil?

    D. Bleach or dish soap?

    E. Lighter fluid or white paint?

    F.  Sweat or blood?

    G. Mercury or chocolate syrup?

    H. I'm getting uncomfortable with how much this is starting to sound like a list of fluids one might find at a crime scene.

ANSWERS:

    A. Cream is more viscous than coffee.

    B. Brandy is more viscous than vodka.

    C. Motor oil is more viscous than gasoline.

D. Dish soap is more viscous than bleach, but bleach is way better at cleaning up incriminating evidence.

E. Paint is more viscous than lighter fluid, but it's probably better to burn evidence than paint over it.

F. Blood is more viscous than sweat. And much more damning.

G. Chocolate syrup is more viscous than mercury and covers its deadly metallic taste very well.

H. So creepy.

**Extra credit: In all but one of the pairs listed, the liquid with higher viscosity also has higher density. Which pair is the exception?**

**ANSWER:** Mercury is denser than chocolate syrup, but the syrup is more viscous. Mercury is a very strange and heavy fluid.

## ((( TRY THIS! )))

Invite a friend over. Open two cans of root beer, scoop vanilla ice cream into tall glasses, and pour the root beer in the glasses over the ice cream. Put a straw and a spoon in each glass and suck down the root beer floats with your friend. Next, place the two cans next to each other with about half an inch between them. Rinse out your straw. Ask your friend what he thinks will happen if you blow air between the cans. He may then ask why you would bother to do such a thing. Don't take this personally. What do you think will happen when fast-moving air rushes between the cans? If you're not sure, ask yourself WWBS? (What Would

Bernoulli Say?—a phrase I repeat to myself often). He would say, "Fast-moving fluid has a lower pressure than the still fluid." Then he would say, "Blow through the straw, already!"

When you blow through the straw between the cans, what happens? The cans move toward each other and smack into one another! Bernoulli was right! The air moving fast between the cans has lower pressure than the air hanging out on the other sides of the cans.

By now your friend may have wandered away with his root beer float.

# 15

# CONTROL YOUR CHAOS
## SECOND LAW OF THERMODYNAMICS

In my senior year of high school, I sometimes stared at the picture of Einstein in our physics book and wondered about him. He had eyes that penetrated the secrets of the universe, a half-smiling mouth ready to explain the thoughts of God, and hair that looked like he'd passed out at a kegger and his friends had jacked it up with egg whites and a blow-dryer. Who was this puzzling genius, and why had he not mastered the complexities of operating a comb? The answers to these questions were coming in Coach Lucido's explanation of the second law of thermodynamics.

By senior year, I'd fully digested and tested the first law of thermodynamics: energy is neither created nor destroyed; it only changes forms. I was an expert at converting the potential energy of elevation into kinetic energy by jumping with my friend Amy out of her second-story window without crushing her mom's zucchini plants. I'd practiced the conversion of elastic potential energy into kinetic energy with many precise rubber band strikes to the back of an unsuspecting head. (Not recommended.)

It was time to graduate from the energy exchanges of the first law of thermodynamics. Coach Lucido delivered the alarming news of the second law: disorder is always increasing. Every process, activity, change, and flicker of the world makes it less ordered. We are bit by bit becoming messier and more chaotic. Even more unsettling, these activities that increase chaos are irreversible. Like an eye shot out with a rubber band, there was no making it right again.

To understand what this irreversibility thing was all about, Coach Lucido invited us to consider a simple energy conversion with which we were all familiar—a log burning in a fireplace. He asked us what we would do if, after a log was burned, we wanted to unburn it. We could try to reverse the burning process by giving the burned log back the heat it had given us. We could even shine some light on it to give back the glow it had given us, but heating and lighting that sad little charred log would not help it renew itself. It would only leave us with a warmed-up, spotlighted, and still very burned log. We would be in the same awkward position that all the king's men were in with dear, dumb Humpty Dumpty—forced to deliver the news to the burned log that we can't put it back together again. Instead of spending too much time running around trying to patch it together, though, our understanding of the second law would allow us to accept the state of affairs and move on. We know by now that the king's men should have been straight with Mr. Dumpty. They should have simply said, "We can't help you out. Anything this broken and scattered can't ever be put back together in quite the same way." (Also, they might have mentioned to Humpty that if you are an egg you shouldn't doze off on a high wall. In fact, you should really stay indoors surrounded by pillows. There are a lot of ways for an egg to get into some serious trouble out there in the chaos-filled world.)

In conversation, it is acceptable, even poetic, to call everyday disorder "entropy." When talking about the second law, though, we are talking about what entropy means to a scientist or engineer. We define entropy in a very specific (and self-serving) way: the unavailability of a system's energy to do work. In other words, entropy is a measure of how much of the system's energy can't be harnessed and used. And there's nothing engineers like to do more than harness thermal energy. It's the core idea behind our beloved machines and power plants. Our first step is to use coal, petroleum fuel, natural gas, or nuclear energy to make steam to spin a turbine. That turbine then spins a generator from which we get electricity. In the early days of converting thermal energy we used steam to power trains, boats, tractors, and even printers. Seriously, we couldn't get enough of steam.

When we say "not available to do work" we are hinting at the difference between concentrated energy and diffused energy. If we had a small metal tank full of steam, that thing is certainly available to do work pushing pistons on a little steam engine. If we open a valve and let that steam leak out into our laboratory for some humidification, that energy from the steam is now diffused, scattered, and unavailable to power anything but a lovely complexion.

Measuring the kinetic energy (motion) of each molecule would be a mind-boggling operation. We use temperature to measure the average energy of the molecules, not the energy of each one. If we look at our friend steam on a molecular level, we would find a bunch of water molecules zipping around and crashing into each other. The surprising part is that even in a tank of steam that is hanging out at one fairly constant temperature, the molecules of steam in it are moving at different speeds. If you were to plot the kinetic energy (a function of speed) of those steam molecules on a graph, it would look a lot like the IQ of all the people in a city. There are a few slow ones, a few really quick ones, and a lot of very average ones. It's impossible to track the motion and energy of every molecule or atom, so we use temperature as a summary of their speed and energy.

## Post-Sauna Entropy

Now that we've decided on a definition of entropy, let's look at an example of how it constantly increases even when you didn't mean to make a chaotic mess. If you are in a sauna getting sleepy and you remember that it's not safe to doze off in there, you will turn off the heat and open the door before nodding off. When you wake up, you will be shivering and clinging to a moist towel. The adjoining room is now a bit warmer and the sauna is a lot cooler. The rooms have settled on the same temperature. In doing so, the energy of the atoms in the room got a boost and sped up, and the energy of the atoms in the sauna dropped and slowed down. Overall, the entropy increased. There is less *useful* energy overall. It's all about us and how we can use the energy for our selfish sauna and steam engine needs.

## Maxwell's Atom-Sorting Demon

In the mid-nineteenth century, James Clerk Maxwell, a brilliant and gloriously bearded Scottish physicist, engaged in a thought experiment to find a situation, if even imaginary, in which entropy could increase. Maxwell imagined a tiny creature sorting out hot (fast) gas molecules from slow (cold) gas molecules. This little being was later named a "demon" by Lord Kelvin, one of Maxwell's frenemies. The demon's job, as Maxwell imagined it, would be to stand at a tiny gate that separated fast gas molecules (hot side) from slow gas molecules (cold side) and open the gate only when an especially slow-moving molecule was headed from the hot to the cold area, or an especially fast-moving molecule headed from the cold to the hot area. The demon would be able to do this because even in a volume of hot gas, there are some slowpoke molecules, and in a volume of cold gas there are some speedy molecules.

So if such a molecule-sorting demon existed, a system's entropy could decrease. The cold side would get colder and the hot side hotter. The demon would whip entropy's chaotic behind. The second law of thermodynamics would be the second suggestion of thermodynamics. There would be parades with banners reading, "Good-bye to increasing chaos! The second law is broken! We are free!"

Before we celebrate our liberation from entropy, let's look again at Maxwell's demon. Something needs to power this freaky little creature so it can toss its microscopic lasso around and sort fast from slow molecules. So now we need to give Mr. Demon a little engine, fuel cell, or bag of nuclear demon snacks. But he won't burn that with perfect efficiency. We know from every other engine, demon, or creation we have fueled, including our own bodies, that nothing runs with perfect efficiency. In a car, for example, while fuel is used to move the wheels, it is also wasted heating up the hood of your car, coughing up warm exhaust, and making a lot of noise. We have no reason to expect any better efficiency from our little demon. So we are still faced with an overall increase in entropy even with the demon's expert help sorting fast high-energy molecules from slow low-energy molecules.

Maxwell died when he was forty-eight, so he didn't have a chance to respond to these comments about his demon, but they wouldn't have bothered him. He created the demon thought experiment only to show that the second law was describing not some literal river of chaos, but the summarized actions of a gazillion molecules zipping around at different speeds. Also, it was cool to have a fictional creature named Maxwell's demon so that he could tell his younger students that it looked like a dragon and he had ridden it to school every day. Kids will believe anything.

## Define a Control Volume

Let's look at a much less scientific entropy analogy. Keep in mind, we are now in very nontechnical territory; I just want to make a point about how to define entropy and how it increases. One could argue that a baby growing in a mother's uterus is a miraculous reversal of nature's tendency to disorder itself. Let's look at this process from the very beginning and see if that's really the case.

An afternoon nap, a long, shared shower during which neither party actually bothers to shave nor shampoo, or a winter
power outage are all good places to start a baby. After that, some
incredible organization takes place. A little person begins taking
shape with a tiny spinal cord, fingers, and a nose that unfortunately looks like Grandpa Edgar's. Doesn't this process defy the
spirit of the second law of thermodynamics? If everything tends
toward disorder, how is this feat of supreme organization taking
place?

If you pull back and look at a larger slice of the world, you'll
see that the overall effect is still one of increasing disorder. Our
mother-to-be is eating food that grew using the sun's energy.
(A lot of food, and nobody better say a word about it—not . . .
a . . . word.) Her body took this food and made baby cells. All of
this is an inefficient process. By the time that little miracle is
taking his first breath and screaming his indignant lungs out
after being kicked out of his warm cocoon, a lot of energy has
been burned in his creation, all of it originating from the most
concentrated source of energy in our lives, the Sun. (Thank goodness the Sun has plenty of energy to spare. It will take billions of
years to burn, scatter, and diffuse all of it.)

## Using Irreversibility to Keep Secrets

It will be worth remembering which processes are irreversible
and which aren't when you are working as a secret agent and are
given printed directions to a government safe house in Turkey.
Once you have memorized the coordinates and researched a
good place nearby to get lamb meatballs and a fig smoothie, do

you shred the instructions or burn them? Which one is a truly irreversible process? Government interns, given enough time in a windowless bunker, can piece together a shredded note. There is no amount of effort, though, that can unburn a piece of paper and restore it to its original form. Even if all the energy given off in the form of heat and light was returned to the note, that piece of paper and its map can't be returned to a readable condition. The energy is burned, scattered, and dispersed in a chaotic way that makes it impossible to reorder. So opt out of shredding. Burn the note quickly in your hotel's bathroom and wash it down the sink before the smoke triggers the sprinklers. You don't need the chaos of a soaked hotel room and strobe alarms; just the right amount of chaos will do.

## The Physics of Life: Reigning over Chaos

That's the key to dealing with this inevitable chaos—cultivating the right amount and type. Focus your efforts on using that irreversibility of disorder in your favor and directing the inevitable chaos into areas that don't hurt you.

Chaos will always increase when energy is being used, but how do you keep it tamed? Must chaos always increase in your life, making it crazier and messier all the time until you have to change your name and leave the country? Not if you look at the fine print of the second law. It applies to an isolated system. You can draw a little dashed line around any system—the universe, the planet, or a single New York City playground. If you decide your *life* is the system, you can direct chaos into places in your life where you don't mind it hanging out for a while.

Some of us (no need to point fingers or state full names) have, during certain periods in our lives, experienced quite a craving for chaos. My theory is that even if we don't know the second law of thermodynamics by name, we instinctively understand the need for chaos to increase in the universe. Maybe we all take turns, for the good of humanity, throwing ourselves onto the pile of disorder to even the score for all of those with safely ordered lives. Whatever the case, actively courting chaos is a dangerous game. Listening to the sirens get louder and louder after throwing an ex's belongings into a well-tended bonfire in the front yard, one becomes philosophical about these things.

I've learned to get my adrenaline fix with bumpy flights in bush planes over Alaskan glaciers, attending the terrifying opening night of my first produced play, or destroying my kitchen trying to make shrimp gumbo for six of my most interesting friends. I know I will have chaos, entropy, drama . . . whatever you want to call it in my life, so I've chosen to give it a seat at my table as an invited guest rather than have it crash in at an inconvenient time and place.

It took me a while to discern between the bad chaos and the good chaos, but I can smell the difference between them now. Dangerous, unproductive chaos smells like a plastic sofa getting burned by an unwatched cigarette. Good chaos smells like a beach thunderstorm drowning out a driftwood fire. They are similar smoky aromas, but they each have their own distinct bouquet. With practice, you can smell the difference a mile away. One is dark and greasy; the other is clean and steamy.

To make sure chaos knows that it is welcome in my life and on my terms, I throw all my gum wrappers, apple cores, and gas receipts on the passenger-side floor of my car. I do it with an intentional flourish to make sure the gods of entropy notice my offering. I allow this small part of my life to be a visible and embarrassing disaster. If someone gets in my car and gives me a look that shows his and her deep disappointment in my inability to keep my car clean, I simply offer up a small, helpless shrug. With that gesture I communicate to my passenger, *Yes, that is the best I can do. It is impossible for me to be more organized than that. Please, have a seat and buckle up. Those empty coffee cups will serve as additional air bags.* I refuse to cheapen my offering to the second law with an insincere apology.

Letting this one small part of my life be a mess allows me to follow the second law of thermodynamics in my own little universe. I've chosen an inconsequential place for entropy to gather, just like Einstein did. He had a whole universe to organize in his head. He was trying to understand how all the forces of the universe were related, how they all originated at the moment of the Big Bang. He wanted to define and quantify every piece in the atomic puzzle. He was literally on a mission to organize our entire universe. That's a lot to define and tame. It's difficult enough to ask the right questions, much less distill the answers into simple, elegant equations. So he let chaos reign in one spot in his life. Without explanation or apology to the rest of the world, he turned over his hair to the gods of entropy. The guy was a genius.

## ((( PHYSICS PRACTICE )))

**1.** After the first and second laws of thermodynamics were in circulation, an assumption that previously didn't have a name was dubbed the zeroth law of thermodynamics. It is this: if two systems are in thermal equilibrium with a third, they are in thermal equilibrium with each other.

At the risk of insulting the zeroth law, which already has kind of an embarrassing name, it seems a little obvious. It states that if two things are like a third thing they are also like each other. Apparently scientists needed to state this to clear a few things up about the nature of heat. Let's think of some other terribly obvious zeroth laws.

A. What should be the zeroth law of motion?

B. How about Murphy's zeroth law?

C. The zeroth law of diminishing returns?

ANSWERS:

A. If something is in motion, it is moving.

B. If something can go wrong, it can also not go wrong.

C. If you don't start an activity, it will never yield returns at all.

# 16

## KNOW WHEN IT'S TIME TO TAKE COVER

WAVES

I should have known, when I was drinking beer with my friends in my senior year of high school,* that I was going to be obsessed with physics for the rest of my life. I insisted that they all blow in the tops of their bottles, then take a drink and blow again to notice how the note was lower because less beer allowed for a longer standing wave in the bottle, which lowered the frequency of the note.

It was hard for them to hear me over the car speakers from Kari's Chevelle, parked with the doors open so we could hear the B-52s. I made sure to speak very loudly when I explained that frequency is inversely proportional to wavelength. When they

---

* If you are under the legal age and thinking, "Hey, she drank in high school and turned out just fine. I'll do that, too," I ask you to reconsider. Drinking is bad for your complexion.

turned up the music, I drew a sine wave in the dust on Kari's car fender to help them visualize how the space in their beer bottle defined the frequency of the note and how amplitude (height of the wave) defines how loud the sound is while frequency (how close together the waves are) defines how low or high the pitch is. I then drew a guitar with a vibrating string to show them how a finger moving up the neck of a guitar shortens the string and makes the note coming from the guitar higher. If they'd bothered to compare my sketches they'd see how air bunching up in a bottle to make standing waves is similar to a guitar string getting shorter and longer. Instead Kari told me to stop it because she didn't want me to permanently scratch her car.

They'd come to expect this kind of thing after I'd consumed exactly one quarter of a beer. Not one of them ever took the time to study my illustrations or equations. So rude. Whatever. Some people just don't appreciate scientific discourse and free tutoring. I still think wave formation and behavior are fascinating.

## The Poetry of Waves

If you throw a pebble into the street, you don't see any waves. If you drop that same pebble in a koi pond, you see waves spread out in circles around the pebble's point of entry. You can then write a sweet little haiku about it:

*Small falling pebble*
*Puckers the blue pond's surface*
*Waves form and transform*

A rock landing in the street would never inspire haiku. Without the water, there is no wave and no poetry because there is nothing for the pebble to push around.

Like a pebble moves water to make waves, a stomp of your foot moves air to make sound waves. Air is sound's poetic medium. A sound-making event such as a beer bottle crashing into pieces on your best friend's driveway pushes air molecules in a wave that then nudges against the inner ear of her father, resulting in that same father standing in the front door in his pajamas and deciding that you are never allowed over to her house ever again unless it is to help with chemistry homework and even that will be closely supervised. If a beer bottle crashes on a driveway and there's no air to carry the sound, no one hears it. But there was air to carry the sound, and now crucial discussions about party locations must henceforth take place furtively over stoichiometry worksheets.

## The Doppler Duck

One of my favorite everyday physics phenomena is the changing pitch of a passing siren or train whistle, named for the sickly but handsome Austrian scientist Christian Doppler. The pitch starts high, and then drops as the car or train passes. The easiest way to visualize what is happening with sound waves when you hear the Doppler effect is to watch ducks in a pond. When a duck paddles through water, the ripples in front of it bunch together and the ones behind it stretch farther apart.

The same thing happens with an ambulance's siren. As it speeds ahead, the sound waves from its siren bunch together in

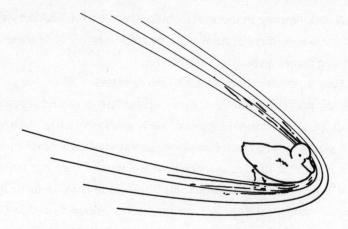

the air and the ones behind it stretch farther apart. When the peaks of sound waves are closer together it hits our ear as a higher frequency. When peaks are farther apart, it greets us as a lower frequency.

By the time Doppler was doing his research, it was known that light, like sound, is a wave.* The frequency of a light wave defines its color, and the amplitude of the wave defines how bright it is. Intuitively, it makes sense that color's pitch and brightness are the equivalents of sound's frequency and volume. The rainbow of colors we can see goes from the lower-frequency red to higher-frequency blue. In the same way that a siren sounds higher coming toward you and lower going away, Doppler theorized that light would change frequency, and therefore color, while it was moving toward or away from an observer. If that were true, galaxies moving away from ours would look red as

---

* Light is also a particle or an energy packet. It refuses to be defined by your labels. It's all, "You don't know me. I'll be whatever I want!" More on that later.

they speed away from our eye. The frequency of their waves would decrease with more space between them, like the ripples behind a paddling duck or the sound blaring from an ambulance heading away from us. This change of frequency is a change of color to our eye. The lower the frequency, the redder it would look. A galaxy speeding toward us would squish its light waves together for higher frequency and look blue.

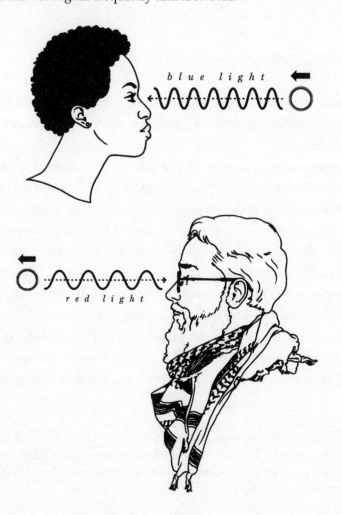

It was easy enough to confirm Doppler's theory with sound, but it took thirty years for Doppler's redshift and blueshift to be confirmed in the case of light. After it was accepted, the Doppler story got even more interesting when it was understood in terms of Einstein's relativity, stretching space-time, and all that stuff that will keep you staring at the ceiling and wondering about the nature of your existence.*

## Racing Waves

Sound travels at 768 miles per hour. Light travels at 186,000 miles per *second*. No contest. In a footrace, light embarrasses sound. It's not even close.

If you are ever a war correspondent covering a growing rebellion in a charming and well-armed city, your knowledge of sound waves could be very useful.

As you are speaking into the camera and answering questions from the anchor with perfectly sprayed hair at the DC desk back in the United States about the developing story and bombs start to fall in the ancient vineyards behind you, the anchor may ask you how far away you are from the bombing. Your first instinct might be to say, "Too #%*^ing close! I'm getting the hell out of here!" But then you remember that viewers are relying on your coolheaded assessment of the situation to help them

---

* If modern physics is new to you, prepare to be totally unsettled or inspired in the last chapters of this book. You may want to sell everything you own, including your shoes, in a giant garage sale and walk barefoot to a monastery after reading about relativity and subatomic particles.

understand the politics of a complex and historic region. If you rock this assignment, you could be on your way to his comfortable non-bomb-adjacent chair in the DC studio. So you tell him you aren't sure how far away the bombing is but will get back to him on that. Then you kill a few seconds with some eloquent words about the history of the region as you scan the vineyard.

When you see the flash of the next bomb, you count the seconds (one one thousand, two one thousand) until you hear the BLAM of the explosion. You know that sound travels about a mile in five seconds and you counted two seconds between the flash and the sound. You do a quick bit of math in your head and calmly tell Mr. Spraytan that you are a little less than half a mile away from the bombing. When the next bomb falls, you count only one second between seeing the light and hearing the KABLANG, you calmly say, "That one was about 400 meters away. We are going to find some cover." Then you make sure that your microphone is off before you yell, "Where the #$%@ is the basement in this place?!"

## ⟨⟨ PHYSICS PRACTICE ⟩⟩

1. You are in a cross-country skiing, crossbow, and drinking competition with an off-duty Finnish coastal jaeger. The winner will get a pile of Euros and a patrol boat. After two laps, you are holding your own in the skiing and crossbow rounds but he's drinking the skinny bottles of gooseberry ale at the end of each stage much faster than you are. You suspect his bottle isn't as full as yours at the start of each round, that maybe his jaeger friends are pre-drinking some of his wine to give him

an edge. Since the bottle's glass is dark, though, you can't see the wine level. How can you check for foul play?

ANSWER: Sprint through the next skiing and shooting stage and get to the wine bottles before Drinky Helsinki does. Before drinking out of your bottle, blow in your bottle and then his. If the note from his bottle is lower, someone has already sipped some of his wine. Call a Finnish foul. He has to take the next lap blindfolded.

2. You have your nieces for the Fourth of July. You thought you had enough snacks to keep a five- and a seven-year-old happy for a few hours, but they have already plowed through jumbo bags of string cheese and animal crackers, and a pallet of juice boxes. (They are so tiny. How do they do it?) Your plan is to bring them to the city's fireworks display, where there are free hot dogs and ice cream for kids. The roads are jammed with people who also heard about the free hot dogs and ice cream. Traffic is moving at roughly 5 miles per hour. You know that if it takes longer than 15 minutes to make it to the hot dogs, you will have a double meltdown on your hands. To decide whether to make a U-turn and head to a drive-thru for dinner or to tough it out and get to the hot dog party, you count 4 seconds between seeing the fireworks and hearing them. If traffic doesn't speed up, how long will it be until you can get your adorable little niece-beasts fed?

ANSWER: Sound travels at 768 miles/hour. Convert that to 0.2133 mile/second and multiply the 4 seconds it takes for you to hear the sound of the fireworks to find that you are 0.85 mile away. (Don't even bother calculating how long it took the light to reach you. It's something like 0.000005 second.) At 5 miles/hour, you'll cover 0.85

mile in 10.2 minutes. You're going to make it. Take deep breaths and remember that your nieces will take care of you when you are old. When that happens, it will be your turn to get all crazy and screechy when you are out of animal crackers. Ah, the beautiful circle of life.

## ⟪ TRY THIS! ⟫

**1. Find a music store and go inside. Pick up a guitar. Pluck a string with one hand and press that string to the fretboard (neck of the guitar) with the other. Now pluck that string and slide your other finger up the string while it is ringing. Does the sound get lower or higher? Why?**

ANSWER: (If you already play guitar, you know the answer. Go ahead and solo madly while I explain this.) When you move your finger up

the string, you shorten the effective length of the string. This shortens the wavelength of the resonating string. Since wavelength and frequency are inversely proportional, frequency has to go up. A higher frequency means you get a higher-sounding note.

**2. While still in the music store, ask to test a microphone and speakers. When the salesperson has wandered away to ring up a tambourine, stand with the microphone in your hand and the speaker at your back or over your shoulder. Sing your loudest, highest note. What happens? Why?**

ANSWERS: Very loud, ever-growing feedback will occur. When you wail into the microphone, your voice is amplified, comes out the speakers, and then goes right back into the microphone, amplified, and on and on until it's PAINFULLY LOUD!

To make it stop, turn the speakers all the way down. If you can't reach the volume knob right away, at least turn the microphone away from the speaker while you look. To understand why feedback occurs, we need only examine the word "amplifier." An amplifier takes sound waves and amplifies them—makes their amplitude bigger. It doesn't do anything to the frequency, just the amplitude. So it makes sound louder and leaves the pitch alone.

# 17

# WATCH A BOILING POT . . . OR DON'T
## PHASE CHANGES

"It's just a phase," my mom assured Chuck when my sister's and my personalities hit that adolescent, erratic stretch. As the instant dad to two teenage girls, Chuck did his best to keep up with our changing tastes and philosophies. One week his record collection was laughable and his clothes were dorky. The next week my sister was playing Chuck's Rolling Stones albums on her stereo and I was wearing his jean jacket to school. That was one advantage of having a young stepfather. He wasn't old enough to be totally embarrassing, like the other dads. Now if we could only get him to shave the mustache. It was acceptable the month before, but was now suddenly *so* "early '70s rookie cop."

Sister Anne Eugene was not impressed with my retro jean jacket. As I walked by her in the inner court her right hand shot out and enclosed on my upper arm with a force that made me suspect she'd suffered a terrible accident and her right side had been reassembled with robotic parts. She finished her conversation with the French teacher while I tried to recall from health class how long a tourniquet could remain on a limb before the lack of blood flow forced an amputation. I made it out of the inner court with both arms that day, but not the jean jacket.

My mom knew that there was no reason to make a fuss about my uniform-breaking outerwear, occasionally green streaked hair, or bottomless appetite for English muffin minipizzas. It was just a phase. I would pass through it to another phase, or come back to a previous phase, but I would still be the same person in the long run. In the meantime I was trying out and discarding ideas, cycling through phases, impatient to be an adult. I couldn't wait to be free. I didn't know exactly what "free" meant, but I hoped it involved a black motorcycle. When all of the quick changes of mood and style got too confusing, Chuck found something to do in the garage, like load shotgun shells or read the *Alaskan Milepost* and circle all the stops with propane filling stations and hot showers.

Like phase-hopping teenagers, elements and compounds can freeze, melt, and vaporize. They change forms from a solid, to a liquid, to a gas, but they are basically the same atoms or compounds no matter what phase they are in. Even the elements we consider reliably solid have their melting and boiling points. Silver melts at 1,763 degrees Fahrenheit (crazy hot) and will boil if you can get it to 3,924 degrees Fahrenheit (way stupidly hot). Notoriously gassy nitrogen will condense into a liquid if you aggressively refrigerate it to minus 346 degrees Fahrenheit. (These temperatures shift a bit if you aren't at sea level atmospheric pressure; more on that later.) None of these changes is permanent. Change the temperature (and pressure) and silver or nitrogen will happily convert to yet another new state.

Water's phase changes occur in temperature ranges that are within daily human experience. We have all seen ice melt to water, and water boil to steam, and the flip side of these changes— freezing and condensing. Don't be confused by all the names for these phase changes. Solids melt into liquids. Liquids boil into a gas. If that liquid happens to be water, when it boils we call the result steam or water vapor. When it changes direction and turns back into a liquid, we say it is condensing, and may call the resulting liquid dew. (Engineers sound oddly like old romantic English poets describing a morning in the countryside when we refer to "dew point" at different conditions.)

## The Steady Temperature of Boiling Water

Since water's phase changes can happen right in front of us in the kitchen at temperatures we can achieve, we have an opportunity to get into water's business and snoop around in its world of phase changes. If you were to stick a thermometer in boiling water, you would notice that no matter how much you turned up the stove, the temperature of the water remains at 212 degrees F (100 degrees C). You can add all the heat you want, but none of it will be used to make any of those water molecules any hotter. The water is using all of the heat energy from the stove to transform into steam. It's an energy-intensive process to make that phase change.

As we discussed in the chapter on buoyancy, water is a tightly knit and organized community of $H_2O$ molecules. They have strong bonds and good family values. Leaving this group is a very big step for a water molecule. To transform into steam, it needs to gather a giant, energetic running start before taking flight. That package of energy required by an atom or molecule to make the transition from liquid to gas is specific to the atom or molecule in question and is called the "enthalpy of evaporation," "enthalpy of vaporization," or "heat of vaporization." All of these describe the energy required to break the intermolecular bonds that liquids have with each other and take off as a gas.

As the water molecule is gathering energy and moving around, getting ready to make its big break, the air molecules on the surface of the water are pushing down with their regular atmospheric pressure. That water molecule needs to overcome the bonds of the other water molecules *and* the air pressure on the surface. Only then is it ready to make the soaring high jump into steamhood.

Now, if your pot of water happens to be in a ski cabin in the mountains of Whistler, British Columbia, you'll be at a higher altitude and there will be fewer air molecules hanging out on the surface of the water. With lower air pressure, it's easier for water molecules to break free and make the transition into steam. Water will boil at a slightly lower temperature in that mountain cabin. And once it starts boiling, the temperature of the water is fixed. So if you're making some pasta after a day of heli-skiing, you'll need to let those noodles cook a little longer than you would at sea level.

Once an $H_2O$ molecule has made its big break from liquid to steam, if it's in a nice, enclosed, steam-only environment, its temperature can climb again. When it does this, it graduates from regular steam to superheated steam. With no more liquid to hog all the energy for its phase change, the steam can now use the heat to actually get hotter. Engineers love superheated steam. It can be packed with more heat, so it has more energy to spin turbine blades and other useful activities. Also, because it's so much hotter than the boiling point of water, it can lose a little energy without turning into liquid again and getting our precious machinery all drippy wet.

I know what you're wondering. You're saying to yourself, "If water molecules only make the change into steam after they've gathered enough heat energy, what's going on with everyday evaporation when water isn't all that hot? Isn't that a phase change into gas without the water actually boiling?" Great question. Wow. You are really turning into a scientist. The answer to your insightful question can be stated in one important word: diversity.

Even in the tightly knit and fairly homogenous community of water molecules, there are molecules that move a bit slower or a bit faster than the other molecules. Remember, if you were to examine how much kinetic energy each molecule had you'd find a bell curve with lower-energy (slower) molecules on one end and higher-energy (faster) molecules on the other. Most molecules will fall in the middle. These faster molecules are a bit more trouble than the average-energy molecules. Their teachers describe them as "unfocused" and "a distraction to others," but they just have a lot of energy. Enough energy, in fact, that when just the slightest amount of heat hits the water community, those special molecules make their big break into steam, even before the rest of the water boils. Who is "unfocused" now, huh? All the other average-energy molecules will never admit that they are jealous about the evaporated molecules' ability to take off like that. No waiting around through an extended boiling period, no long good-byes, just a backflip into the air and they are vapor! (There is a lot of jostling around at the molecular level, so individual molecules of water slow down and speed up all the time, changing identities between the high- energy molecules and the low-energy molecules.)

This phase change of water molecules is the reason you shiver when you first step out of the shower. As you grope for a towel, those high-energy water molecules are changing phase from liquid to steam. They are evaporating. Evaporation is only different from boiling in that the entire liquid isn't undergoing the change, just the faster molecules on the surface. They need a little chunk of energy to change into a gas, so they take it from your skin. This leaves the slower molecules behind. Since "slower" means lower energy, that registers to our skin as colder. Your body is the stove now, the provider of energy for water molecules to change into steam. They are stealing heat from you to make their big break, leaving you cold and naked. Kind of rude, but that's how those fast molecules are.

Rubbing alcohol feels even colder on your skin than water. Alcohol molecules are not as comfortably bonded to each other as water molecules are, so alcohol changes from a liquid to a gas more easily. While water molecules have stronger bonds, better family values, and more community accountability, alcohol is a whole town of high-energy evaporating rebels by comparison. And when it does evaporate, it requires a larger packet of energy to make the phase change from liquid into gas. With all those alcohol molecules easily taking off and using a fistful of energy on their way, their collective "So long, suckers!" leaves a cooling feel on your skin when you rub alcohol on your arm or spill a shot of vodka in your lap.

## Cooling Off Old-School Style

The systems we use these days to cool down a crowded gym or a computer-filled office are fancy compared to the cooling systems used by ancient people. But many of them use the same concept: evaporative cooling. The key word "evaporative" tips you off to a phase change—liquid to gas, right? To do that, liquid needs a bit of energy. An evaporative system forces a liquid to steal energy from the air to make that jump from liquid to gas, cooling down the air.

So on a hot day in ancient times, clever Persians or Indians would hang a wet grass mat over their doorway. This gave hot air something to spend its energy on. As a wind blew the air through the mat, the water in the mat evaporated, taking little chunks of energy out of the air for its phase change. The air coming into the house would be a bit cooler. Maybe those ancient overheated people didn't know exactly what was going on at the molecular level, but they knew that their homes were more comfortable. Artwork from Egypt shows servants fanning clay jars of water. The moving air would encourage evaporation of the molecules on the surface and leave cooler water behind. The fanning would also help sweat evaporate on the skin of the people sitting in the room. Everyone was happy . . . except for the servant, whose shoulders were seriously cramping up.

## The Physics of Life: Personal Phase Changes

The phase change from adolescent to adult takes just a few years, but it feels like decades. Like every other high school junior, I was pretty sure I was smarter than my parents. It didn't matter that Chuck had been in Vietnam, Cambodia, Israel, and knew how to speak Spanish and a bit of Hebrew. It didn't matter that my mom had been all over the country as a flight attendant and lived on her own in San Francisco when the beat poets were snapping applause in coffee shops. I still internally rolled my eyes when either of them gave me advice. I was looking forward to the day when they would acknowledge that I was indeed wiser and cooler than they. Maybe we could get a cake that said something to that effect in cursive frosting.

When I told my mom that I was dying to go to college, have my own dorm room, and wear jeans every day to class, she told me it would happen before I knew it. In the meantime, like generations of teenagers before me, I accepted the help my parents gave me without much thanks. Mom would type my papers, correcting spelling and adding semicolons. Chuck would get up before it was light and drive an hour to his job as a diesel mechanic and return long after I had eaten dinner. He would then hand his paycheck over every two weeks to my mom to pay for mortgage, bills, and my high school tuition. The weekends he didn't work, he would sometimes lay on the living room floor with a heating pad or ice pack on his back.

At our round oak kitchen table Chuck would eat his late dinner and look at his books while I did my homework. After dinner he would study with me, flipping through his professional engineering licensing exam study guide that was covered in his diesel fingerprints and coffee stains. I was focused on finishing my homework as quickly as I could to make some important phone calls and find out which track girls were dating football boys. (Those mixed relationships never worked.)

One night at the kitchen table Chuck asked me about something he was reading. He didn't understand the math describing a thermodynamics cycle. I didn't know the cycle, but I knew the math. I showed him how to cross-multiply, cancel, and isolate the answer. He followed along. So far, he'd been the one doing the teaching. He made it his job to show me the crucial life skills—how to fish, operate a drill, and deliver an elbow strike to the throat. It felt strange to teach him something.

Sitting next to him, watching him write out the numbers, I looked at his hands and realized that the grease covered a lot of scars until you studied them up close. Some nights, my mom would use tweezers to pull metal slivers from his hands after dinner. For the first time, I was worried for him. I was used to worrying about my mom because of her seizures, but now I worried about her and Chuck in a different way. I wondered how they would get by when Chuck was too old or his back was too sore to climb under diesel generators.

Before Chuck married my mom and adopted my sister and me, we lived in *that* house on the block. The one with the police or ambulance in the driveway because my sister or I had called them when mom had a seizure and we didn't know what to do. We were the kids who went next door to borrow eggs, saying we were making cookies when we were really scrambling them for dinner. We lived in the house that the social worker visited to check our kitchen and make sure we had pots and pans so we could use our food stamps properly. Chuck changed all that. We now lived in a quiet little house with honeysuckle crawling up the fence and enjoyed a life that he carried on his shoulders. He was only thirty-three, but for the first time, he looked tired to me. Watching him write out a quadratic equation at our oak table with his scarred hands, I realized he couldn't carry us forever.

I would have to carry myself. I might even have to carry Chuck and my mom. My sister had taken a quick stab at college but didn't stay. It was up to me. I knew what I had to do. I would get my degree first and then support Mom and Chuck while he got his engineering degree.

That's when my phase change from girl to adult began. I didn't shake free of my parents like I thought I would when I turned into an adult. Instead, I understood my responsibility to them. I was passing them by in education. I needed to use that privilege to help them. I understood why I needed to get good grades, go to college, and choose a major that would get me a real job. They would never ask me to support them, but in that moment, when I started changing into an adult, I knew I wanted to be ready to do just that.

As with the temperature of boiling water, nothing changed right away that you could measure. Mom and Chuck still wrote checks for my high school tuition, then my college tuition. I still showed them report cards. I still internally rolled my eyes when they gave me advice. But now I knew: They needed me. I had to be smart and I had to succeed.

So near the end of my junior year of high school, I stopped wishing I was grown up. I was no longer staring at the pot, waiting for it to boil. I applied to college and researched starting salaries for different four-year degrees. As it always happens, when I stopped watching and waiting, the water was boiling like crazy.

## ((( PHYSICS PRACTICE )))

**1. If you have cold liquid nitrogen in a tank and you want to keep it in a liquid state as long as possible as it warms up, should you lower or raise the pressure in the tank?**

ANSWER: High pressure on a liquid makes it harder for molecules to change from a liquid to a gas. Faced with higher pressure, molecules need more energy to launch into the gas state. In other words, they need a higher temperature to boil. So keeping liquid nitrogen in a highly pressurized system will raise its boiling temperature, making it easier to keep in a liquid state. Raise that pressure!

**2. Now that you understand how water evaporates on the skin, explain why a dry heat is so much more comfortable than humid heat. What's going on when you sweat?**

ANSWER: When it's hot, you sweat. Your sweat evaporates into the surrounding air. That evaporation process steals a bit of heat from your body and cools you down. Dry air absorbs water vapor more readily than air that is already crammed full of water condensate (aka humidity). Because of all this, the cooling effect of sweat is more dramatic in dry air than in "wet" air. Also, humidity frizzes your hair, and that makes you feel frumpy. When you feel frumpy, a hot day is hotter. Scientific fact. Look it up.

## ((( TRY THIS! )))

If you head out on a walkabout in Australia with every intention of enjoying a spiritual sabbatical and find yourself lost, you will want to know how to use water's phase changes to make yourself some clean water. Best to practice now while you're not crazy with thirst and staring at a rock formation wondering why it looks like an angry cockapoo. Let's give it a shot:

With a big bowl, a little bowl, a little rock, and some plastic wrap, this project is really easy. Just put a few inches of the dirty water in the big bowl, put the little bowl inside the big bowl (also facing up), and then stretch the plastic wrap over the big bowl. Place the rock on the plastic wrap, right in the middle. Put the whole rig in sunlight. The water will evaporate, cling to the plastic wrap, and run down to the low point in the middle and drip into the small bowl. When it evaporates, it leaves the dirt behind and leaves you with a delicious drink of clean water.

Now you know how to make a small water distiller with the two bowls and the plastic wrap but the real challenge is to make one with less obvious materials. How about a piece of plexiglass and an old bathtub? All you need is a container to hold the dirty water, a flexible surface on which the water can condense, and a place for it to run down that surface and collect.

Extra credit: If there is gasoline in your original dirty water, will this evaporative distillation method result in clean drinking water?

**ANSWER:** Nope. Evaporating water will leave behind dirt and salt, but since gasoline boils even sooner than water, it would evaporate and drip right down into your drinking water collection. This evaporation method only works to separate your water from stuff that boils after water does or doesn't boil at all.

# 18

## USE YOUR POWER WISELY
### ELECTRICITY AND MAGNETISM

By the time Coach Lucido explained electric current to us, I was getting the feeling that every little thing I experienced could be understood using the laws of physics. Life was one big physics laboratory. At the root of every twitch and tic of the universe were the same defining concepts—a force acts on matter (made of atoms), and that matter responds. And even though it wasn't a huge stretch to assume that electric current had something to do with the atom's electrons, I still felt very smart and scientific for guessing correctly before Coach sketched little electrons in motion on the board.

Those electrons (which have a negative charge) hang around the nucleus of atoms (which have a positive charge) in their normal way unless there is some kind of incentive for them to start hopping from atom to atom. One of those incentives invented by us clever humans is the battery. In your car battery there is a plate of electron-rich material on one side and a plate of electron-poor material on the other side. When those sides are connected, electrons rush through a fluid in the battery from the electron-rich side to the electron-poor side. They can't help stampeding from one side to the other. Electrons just aren't themselves when they sense a vacancy around a group of positively charged protons. They shove their way past each other like holiday shoppers desperate for the last Hello Kitty hot air popcorn popper. All this rushing in one direction creates electrical current.

Wires for conducting electricity are most often made out of copper, the atomic champion of electrical conductivity. Copper has 29 protons and 29 electrons; 28 of those electrons are nestled comfortably together in orbits (or, more correctly, energy levels) around the protons and neutrons in the center of the atom. The 29th electron is the perpetual houseguest, not quite a member of the closely knit atomic family. In a copper wire, that black sheep electron, with no particular atomic loyalty, will bounce from atom to atom. There is always another hopping electron right behind it to take its place in the guest orbital. This ease of electron hopping makes copper an excellent conductor.

Before I knew about all of this shoving and hopping of electrons, I assumed the flow of electricity was like the flow of water—a rush of glittery electricity (whatever that is) surging through a wire after we turn on a light switch. But turning on a light switch isn't like turning on a faucet.

When you turn on the light switch, what you are really doing is connecting a wire to a source of atom-hopping electrons, aka electric current. The electrons in the wire then knock into each other in a general direction toward the lightbulb. The electrons don't actually move terribly fast, nor do they move in a straight line. But there are so many electrons crammed in that wire that even a slow pace means plenty of electrons are in motion and making it to the lightbulb.

Electrons do not flow like water; they flow more like a herd of turtles.* Say you were looking at the south end of a long, narrow bridge that connected a turtle mating island to the mainland. The mating season kicked off at exactly 6:00 a.m. (these are very precise turtles). If at 6:00:01 you saw turtles pouring off of that bridge to the island, you'd think that turtles were impressively fast. They sprinted across that bridge in one hundredth of a second! Your confusion about the speed of turtles could be forgiven, because you didn't realize that the bridge had been full of turtles to begin with. They didn't start on the mainland; they were already packed onto the bridge. The same is true with electrons. We give them credit for being quick when really they are jammed tightly in a wire and ready to move at the slightest nudge. That copper wire full of loner electrons runs right up to the lightbulb in our reading lamp, so when we turn on the switch, the electrons closest to the bulb don't have far to go. Instant current and instant light!

To get a sense of exactly how full of electrons that wire is, 1 amp of current represents 6,250,000,000,000,000,000 electrons passing per second. That's why engineers describe current in amps and not numbers of electrons. When we describe quintillions of anything, there are a lot of zeros involved. Engineers don't like strings of zeros. Our calculators are very dear to us and we don't want to burden or bore them with fifteen zeros in a row. I realize it's hard to understand the relationship between an

---

* What? A group of turtles is called a bale, not a herd? Well, that's just ridiculous.

engineer and his or her calculator. Remember, there is no defining or containing love. If you see engineers nuzzling their calculators, just be happy for them. Nothing about their relationship threatens your relationship.

## Electromagnetic Field

Before Coach legitimized it for me, "electromagnetic field" sounded like a phrase that only a cartoon villain poised to take over the planet would use. As it turns out, though, "electromagnetic field" is a real term, and a very necessary one. If we talk about electrons hopping in a direction that creates an electric current, and then try to have a separate discussion about magnetism, we will find ourselves tripping over our scientific feet. Electric and magnetic influences are intimately related. They are parent and child of each other in a circular Zen relationship worthy of a monk's meditation time. If we think of each as a "field," they can be easier to understand.

A charged particle (a proton or an electron, for example) has a field, or area of influence. If another negatively charged particle tried to float into the electrical field of an electron, that particle would be very rudely pushed away. Since the particles didn't actually touch each other, it makes more sense for us to say that their fields (not the particles) interacted with each other. And, of course, the closer the particles get, the more forcefully their fields react to one another. In the same way, magnets have fields. If you move two magnets close to each other, you will feel the pull toward or push away from each other, depending on whether you are nudging opposite or similar poles together.

Here's where it gets interesting: if you move a wire between two magnets through the magnetic field between them, electrons will start to flow in that wire. On the flip side, electrons flowing in a wire will create a magnetic field. Moving magnetic fields and moving electrons are yin and yang, action and reaction to one another. It gets very one-hand-clapping-in-a-forest-with-no-one-to-hear when you start trying to distinguish electric fields from magnetic fields. They are one, Grasshopper. (I bow deeply to your growing wisdom.)

But since humans are notoriously impatient with Zen riddles, instead of pondering the relationships among electric fields, magnetic fields, current, and the nature of life, we focus instead on finding a way to use a stream of electrons to heat up our waffle irons. There are a number of different ways electrical generators take advantage of this magnetic field and current relationship to power your crucial breakfast-making activities. Here's the model for a basic generator. It's so basic that you could re-create it in a pinch.

A bunch of wire is coiled on a frame, looking much like a big, flat beater attachment for a pastry mixer. That beater is then

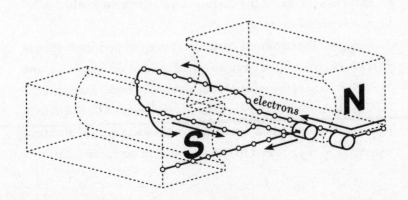

turned between a north magnet and a south magnet that are not moving. The force to turn those coils between the magnets can come from a waterfall, a bicycle, or a steam turbine. Because you have wire moving through a magnetic field, what do you get? That's right, you get current. Easy!

Well, not exactly that easy. There are a few details to figure out once you get the current pulsing through your wire. In the simple generator described above, the current goes way up, then down to 0, then reverses, and then back up again. This is because the relationship between a magnetic field and current has one small quirk. To induce current, the magnetic field has to be at right angles to the movement of the wire. One way to think of it is "against the grain." If the wire is moving with the grain of the magnetic field, the electrons don't get excited and nothing happens. If the wire moves against the grain, the electrons respond and we get current. So in our generator example, the blade of our beater cuts the magnetic current against the grain, and then with the grain. As a result, the current will surge on and off and reverse with each cycle. To make the current run in one direction all the time, early inventors placed a little ring with a space in it around the end of the coil. With that and some contact brushes on the wires receiving the current, they managed to always deliver the current in one direction.

Now that the current is created, we need to deliver it and use it. The most basic bit of electrical wisdom that you need to know to do this is that power equals current multiplied by voltage: $P = I \times V$. (I know it's weird that current is represented by the letter I. Apparently it was called "intensity" in the early days of electrical research.) So if you need to deliver a gob of power to a city, you can do it by increasing either the current or the voltage. But houses take a standard 280 volts, so that's what we have to deliver. If we serve up the power any other way, it's not usable by our radios and blow-dryers. So jack up the current and deliver lots of power! Done!

Well, we can't do that either. If you turn up the current, we won't be able to jam it through the wires delivering the power. Despite what I said earlier about current not flowing like water, delivering current through a wire is similar to sending water through a hose. The longer a hose is, the more it resists the flow of water. The bigger the cross section of the hose, the easier it is for water to flow through it. When we try to deliver current over a long distance, we have a problem: the resistance of the wire carrying our current increases with the length of the wire. We can help the current flow by increasing the cross section of the wire. This would be awfully expensive and we'd need a crew of superhuman lumberjacks to install some very fat, heavy wire. Neither is practical, so we are stuck. How do we deliver lots of power long distances?

## Edison and Tesla Get Ugly

This question of how to deliver power long distances sparked one of the nastiest clashes of ideas in modern history. It's a story filled with big egos, electrocuted dogs, and true love with a pigeon.

Remember Thomas Edison? The great American inventor we learned so much about in school? Well, it turns out they left out the part about him being kind of a jerk. He tried very hard to kill the brilliant idea that made long-distance electric power delivery possible. When you consider how much we rely on long-distance transmission to electrify everything in our life now, it's amazing how Edison was so clearly on the wrong side of history on this one. In the 1880s, when corsets and bustles were hip, and homes were still lit with coal and gas, Edison started a business selling direct current. He would set up a DC generator (like the one described earlier) and then send power to the square mile around it. He served only that small plot of homes and businesses because the resistance of long stretches of wire made it impossible to send a usable amount of power very far.

Then Nikola Tesla showed up from Europe wanting to work for Edison. Tesla had a wild idea. What if we built a power plant, say, right next to Niagara Falls and used the force from the water to spin a generator's rotor and create a kind of power that could be sent long distances? We could power not just sleepy little Buffalo, but also big New York City? How cool would *that* be?

Edison said something like, "That'll never work." Or maybe it was more like, "Who are you to tell me about electricity? I invented this sh*t. Also, your accent is funny and I don't like your jacket."* Why wouldn't he ask Tesla to show him the plans right away? There are two theories:

1. Edison was looking out for his DC-centered empire.
2. Edison didn't know what the hell Tesla was talking about because Tesla was way ahead of him.

---

* Maybe he didn't say all of that. I'm guessing here.

Tesla did work for Edison briefly, but Edison wasn't interested in Tesla's great new idea. Eventually Tesla found himself digging ditches for a winter. (And you thought your first job didn't take full advantage of your special talents!) While he blistered his hands, Tesla continued to mentally piece together his idea. Instead of direct current, he wanted to create alternating current. In a DC generator, brushes and contacts correct for the shifting back and forth of the current as the coil of wire rotated between the generator's magnets. Tesla's idea was this: Let the current shift back and forth. Let it alternate. A lightbulb doesn't know the difference. It just sees electrons flying by. Current is current. The lightbulb doesn't care which direction it's going. It changes too fast for the bulb to go dark.

Tesla knew that this alternating current could be sent long distances while Edison's direct current could not. That's because AC can be sent at a very high voltage (which would deliver lots of power) and then stepped down to the usable 120 volts. Delivering power at high voltage would allow him to use a small current for the long-distance runs. The lower current means small wires could be used.

This "stepping up" the voltage and then "stepping down" again before delivering the power to the city's distribution is a trick you can perform only with alternating current. A surprisingly simple piece of equipment is used to transform the current from one voltage to another voltage. Even its name is simple. It's called a transformer. A step-down transformer steps voltage down from high to low. A step-up transformer steps voltage up from low to high. This transformer is nothing but two big coils of wire that don't touch, but are wound around a shared iron core. Without any current running through one of the coils, they just stare at each other in silence. When an alternating current hits one of the coils and zips back and forth in it, the action begins. What do we know about a moving current? It creates a changing magnetic field. And what does a magnetic field do? It induces current. When one coil of wire has alternating current buzzing back and forth in it, the other coil in the transformer feels that changing magnetic field and starts a current of its own. By setting up the coils so that one is longer than the other, with more wire in its winding, we can step the voltage up or down.

If you try using a voltage transformer with direct current, the current will zip through one coil and do nothing to the other. Direct current doesn't appreciate a transformer. It just sees the transformer as globs of copper wire around a core, not as a brilliant thrumming voltage/current changer. There is no reason for the other coil to react because the current isn't changing direction and creating a field that induces more current. That magical yin and yang thing doesn't occur with direct current. But with AC, it's *on*. The jiggy dancing of electrons in one coil creates a magnetic field that makes the other coil dance. So if you want to send 240 volts to San Jose from a power plant in San Diego, it's best to send all that power at a high voltage and low current through a manageable-size copper wire. When it arrives at a substation in San Jose, use a transformer to "step down" the voltage before sending it into homes.

So Tesla wins, right? We use his alternating current system for our modern power distribution. Yes, but he didn't win soon enough to enjoy the victory. We'll revisit him, but first I want to make sure you know how to use these ideas to not only survive, but also to become the unchallenged ruler of a small neighborhood fiefdom if there is a prolonged power outage in your city.

## Your Firm but Fair Reign as Lord of the Block

What if any piece of this beautifully conceived alternating current system hiccups? What if a solar ray zaps all the transformers and they all stop stepping up and down our voltages? Disaster preparedness experts will be more than happy to tell you what will happen next: Food shortages and then looting, followed by your standard lawless rioting and the collapse of the government.

There is no way to be completely ready for the end of civilized behavior and cable television, but you can at least be one of the people with thirty days of food and water stored in your basement. Also, you'll need a way to protect it (whatever you are comfortable with, but remember, hungry people are crazy). While you are getting prepared, why not also collect the material you need to make a small human-powered generator? After a hard day of building barbed wire fences around your street, you'll be craving a warm cup of Earl Grey. Now that you know the basics of generator operation, you'll be able to do a bit more research and cobble one together with magnets, wire, and a stationary bicycle. Do it before the power goes out so that you can check the Internet for instructions.

Also, while you're preparing for a possible collapse of the power grid, go ahead and practice ways to make a fire after all the matches have run out and the lighters are used up. A small stash of steel wool and a 9-volt battery will be all you need to start a decent fire. Pull the steel wool apart a bit so that it has plenty of air around the fine wires and brush the positive and negative ends of the little square battery against a small nest of steel wool. It will burst into a nice ball of flame as the teensy wires do their best to conduct the current from the battery. The wires conduct more current than they can handle, heat way up, glow red hot, and burn like a tiny campfire.

With a small human-powered generator and a stash of unlikely fire-making materials, you have now positioned yourself to be a very important neighborhood figure in the event of a prolonged power outage. When it happens, you will get an unspoken promotion from Fourth of July party coordinator to Lord of the Block.

Since you are a reasonable person, you will not use your new power to address old grievances . . . unless you feel like it.

## Edison and Tesla's Story Continues

When we left our dear and delicate Tesla, he was digging ditches. When he wasn't digging, he was looking for an investor for his alternating current generator. After many false starts, Tesla found a partner in George Westinghouse. George was a good guy, an inventor, and a believer in AC current. Tesla signed a deal with Westinghouse to create and sell AC current. Edison struck back by electrocuting dogs and horses publicly with AC current to try to convince people it was too dangerous for use. Edison's supporter Harold Brown helped the state of New York execute a prisoner on death row with AC. In a brilliant stroke of negative branding, they called it "Westinghousing" the prisoner.

Westinghouse and Tesla delivered a solid return punch in this war of currents when they won the bid to light up the 1893 World's Columbian Exposition in Chicago. They made quite a show of it, with President Grover Cleveland flipping the switch to fire up the many thousands of bulbs in the "City of Lights." Edison looked on from his little "History of the Lightbulb" booth at the exposition and seethed. (I'm sure Edison showed off something more exciting than that, but it's hard to tell. There is conflicting information over what he had ready for the exposition and what he displayed.)

After that, Tesla and Westinghouse struck another painful blow to Edison. They won the contract to build a power plant at Niagara Falls. They were on a roll, despite Edison's publicity team continuing to "Westinghouse" poodles and ponies onstage. After this run of success, though, Tesla's progress was uneven. When the Westinghouse Company was in a pinch and couldn't pay Tesla his AC royalties, Tesla generously tore up the contract. He didn't want to bankrupt his friend and supporter. It was awfully nice of Tesla, but without that income he struggled to find investors for further research.

When Italian inventor Guglielmo Marconi transmitted the first radio signals across the Atlantic, Tesla pointed out that Marconi used Tesla's patents to achieve this, but was not heard above the cheering. Then one of Tesla's strongest financial supporters died on the *Titanic*. After this run of bad luck, Tesla displayed moments of brilliance and others of downright madness. He counted his steps in threes, couldn't bear to shake hands with people because of germs, and believed beings from another planet were contacting him at his Colorado laboratory.

And then there were the pigeons. When he moved back to New York, he enjoyed feeding them. That's not so strange. It's when he started letting them in his hotel room that it got weird. In his later years, after the stress of fighting Edison and losing support for his projects, Tesla was kicked out of hotel after hotel for not paying (and for the aforementioned pigeons). When his favorite, beloved female pigeon died, Tesla gave up.

Edison spent his last years rich and famous. Tesla died alone, and his body was discovered by a hotel maid. Beyond the personal tragedy, this is a story of wasted brainpower. Tesla had to fight for funding and recognition while Edison did his best to crush him. Tesla had so many other ideas, including wireless (free) power. Sure, he also believed that beings from other planets were sending him messages, and he loved a pigeon like a wife, but we all get a few passes, especially if we invent something that changes the course of history.

Imagine what could have been accomplished if they'd been friendly partners: Edison sitting in his shop with Tesla, working late into the night as they both liked to do. Thomas with his endless patience for trying a thousand possible solutions and Nikola with his intricately designed inventions already detailed in his mind. If they had worked together, what would we have gained? Wireless communication decades earlier? Power plants without emissions or nuclear waste? What did we miss because Edison didn't play nicely with Tesla?

I hope that I can use Edison and Tesla's story to remember that no matter how smart I think I am, someone will be smarter. And that person might be someone who is dressed funny, has a weird accent, doesn't have a lot of friends, and has a compulsion to wipe down all his clean silverware three times with three separate white napkins.

In honoring Tesla, I will always try to remember myself in my first week of high school with my bony knees and stiff skirt, and with rows of desks on either side of me filled with girls I didn't know. When Sister Rochelle asked us to pair up for a volume-measuring exercise, the other girls turned to the friends they had known since kindergarten. I froze. What if there was an odd number of girls and I was the only one without a partner? The girl in front of me swung around and flopped her folded arms on my desk, "Wanna be partners? I hope you know how to measure cubes and densities or whatever because I am seriously lost."

Saved.

## ((( PHYSICS PRACTICE )))

1. Ohm's law can be used to describe the simple electrical system of a battery powering a lightbulb:

Voltage equals current multiplied by resistance.

$V = I \times R$

A. What causes the resistance in this small electrical system?

B. What supplies the voltage?

C. What law have you discovered that should be named after you?

ANSWERS:

A. The lightbulb is the resistance in the circuit. Also, the wire has a little bit of resistance.

B. The battery.

C. Describe your own law with your name. As an example, Miss McKinley's law is this:

$Y = 10 \times (N/S)$ where Y is the number of years it takes to be appreciated for your ideas, N is how new the idea is compared to what is already in use on a scale of 1 to 10, and S is your level of social skills and personal charisma on a scale of 1 to 10.

If your ideas are not very original and you are very charming, you'll be recognized quickly. If your ideas are cutting-edge and hard to understand, but you are socially adept, it'll take a while to be recognized, but it will happen in your lifetime. If your ideas are very new and you are awkward and compulsive, it will take many years for the world to appreciate you. Sorry, Tesla. We appreciate you now, buddy. It would have helped if you weren't so quirky. Still, thanks for the alternating current.

2. **Semiconductor factories require ultrapure water for their process. After sending the water through equipment that removes minerals and other tiny bits of material, sensors tell the plant engineers the resistivity of the water. Which is cleaner, water with a resistance of 18 megaohms, or water with a resistance of 5 mega-ohms?**

ANSWER: The minerals in water make it good at transferring current (more free electrons to nudge each other). When the minerals are removed from the water it's no longer as good at conducting electricity. To state it in the opposite way, pure water is better at resisting current. So high resistivity means the water has fewer minerals in it. Since ohms is a measure of resistance, the 18 mega-ohm water is cleaner than the 5 mega-ohm water.

# ⫷ TRY THIS! ⫸

I know this sounds like a terrible idea, but try using your tongue to complete a circuit. Not on something dangerous like a car battery. Make your own super-low-voltage battery like this: wash a penny and a dime, cut a thick slice from a lemon, and stick the penny and the dime through the lemon slice so that they aren't touching each other. Touch your tongue to the coins. Do you feel a tingling? That's electrical current. Why does this happen?

ANSWER: The acidic lemon juice is ripping electrons from the copper penny and adding them to the dime. This movement of electrons creates a current when your tongue completes the circuit. There are other versions of this penny/dime arrangement. You can stack the pennies and dimes with bits of vinegar-soaked paper towel between them to create a current strong enough to light a tiny bulb. If the tiny tingle wasn't enough excitement, you can touch the connections of a 9-volt battery to your tongue for a nice zing. This is a common activity during sound check when a guitarist is troubleshooting her tuner or pedal and needs to know if the battery in it is dead. A tongue buzz means there's still life in the battery and that the problem with the tuner or pedal is elsewhere. Note: don't ever imply that the problem is the sound person's fault. He will make you pay for an accusation like that all night by putting a gob of confusing reverb in your monitors.

# 19

# CULTIVATE MYSTERY
## THE ELUSIVE ELECTRON

With two weeks left of high school, we were poised to join the grand old traditions of singing our school song for the last time at graduation, burning our plaid skirts behind the tennis courts, and toilet-papering the front yards of boys we liked but had been too afraid to speak to for the past four years. Yes, we had matured into young women our parents and teachers could be truly proud of. Then, in the middle of all this dependable structure, Coach upturned a whole cart of Newton's apples by tearing up the old model of the atom. It wasn't like he slightly changed the wording in one of the laws of thermodynamics or told us that Newton had cheated on his final exams. We could have handled that. But the model of the atom was sacred. It's how we understood bonding, electric current, and the gas laws. Also, it was cute. Our atom looked like a little planet with perfect tiny electron moons.

But electrons don't orbit the nucleus like little moons. "We don't really know what they do," Coach said as he erased the electrons orbiting the atom on the board. He drew fuzzy little chalk clouds around the nucleus. He explained that those clouds are where an electron might be if we went looking for it. As it turned out, electrons don't readily give up their secrets; in fact, they are maddeningly coy.

## The Electron's Debutante Ball

How was the electron introduced to society? How did scientists get from thinking of electricity as a jolt of lightning or a stream of current to thinking of it as tiny packets of charge called electrons, anyway?

It started with party tricks. In the mid-nineteenth century, scientists got a nice "oooo, aaaaah" from their lecture audiences by running a current through an evacuated glass tube and showing off the pretty glow. These tubes evolved into neon signs and are not a big deal to us now, but remember back then people had never seen anything like this and there weren't a lot of entertainment options available. For a wild time, they would read the list of unapproved sexual acts in Leviticus and were tickled to get one orange in their Christmas stocking. Different times.

The more air these show-off scientists vacuumed out of the tube, the nicer it glowed. They would save this glowing tube bit for their encore because they didn't want to field any questions about it. If pressed, they'd have to admit they had no idea why the tube glowed. Many of them had theories, but only the English physicist J. J. Thomson had the guts to announce that he believed the glow was due to some small part of the atom. *Part* of the atom? What? That was quite a claim at the time. There were no *parts* to the atom. Nothing was smaller than the atom. The atom was indivisible, right? Further, J.J. claimed that this smaller-than-an-atom thing had a negative charge. He proved, in a series of experiments, that he was not sniffing chemicals in lab; this negatively charged bit of the atom was real.

But J.J. didn't know how those little negative charges were arranged in the atom and he didn't offer a very elegant model. He suggested that electrons were stuck in a positively charged blob. J.J. called this the plum pudding model. (Plum pudding doesn't actually have any plums, but is more of a cake with raisins in it—distributed randomly, like J.J. imagined the electrons in an atom.) A student of Thomson's, Ernest Rutherford, a farm kid from New Zealand, proved that the atom wasn't a fruitcake-type blob. Rutherford showed that the atom had a positively charged core with the tiny electrons flying around it.

And that's where most of us leave it today. When the electron was examined more closely, though, Rutherford's model (and our understanding of matter) took a strange left turn.

## Courting the Electron

A group of scientists at the beginning of the twentieth century tripped all over each other pursuing the electron. They started with a lot of personal questions. Do you orbit neatly around the nucleus? Are you a particle or a wave? If you were a tree, what kind would you be? The electron gave them just enough information to keep them interested. She sometimes behaved like a wave, other times like a particle. She hovered in a cloud of possibility, but didn't really hover anywhere at all because she wasn't *in* the cloud somewhere, she *was* the cloud, but only if you specifically looked for her there, and even then the act of looking for her would change her whole story. She was like that really interesting girl at a party with an accent that you can't quite place. You never know when she'll arrive, and when she does get there you don't know what the hell she's talking about and then she's gone—vanished into the night with her party dress flowing behind her, leaving only the scent of jasmine and danger.

Despite (or because of) this inconsiderate treatment from the electron, the scientists continued pursuing her. They asked more questions. How fast was she going? Where did she like to hang out? The electron evaded them so successfully that Werner Heisenberg, a hotshot scientist with a rakish pompadour who was not used to being rebuffed, created the uncertainty principle just for her. He announced that if you found either the momentum or position of an electron accurately, you could not find the other. (Of course, you remember that momentum is mass multiplied by velocity and it has a particular direction.)

To get a sense of how strange the uncertainty principle is, imagine driving around in bumper cars with your friends at a carnival. If you could only know either the position or the momentum of their little cars, but not both, you'd have no clue what was going on. Oh, there's Scott in his blue car! You can see where he is, but you have no idea how fast he's going, how big his car is, or in what direction he's headed. But there he is. Hi, Scott! Wait a minute, now you know his mass and how quickly he's headed toward you but you can't tell where he is. Five yards away, maybe? SMACK! Nope, not five yards away. He was right in front of you.

Heisenberg insisted it wasn't that instruments were not accurate enough to find position and momentum at the same time. He determined that it was not in the electron's nature to *have* a certain position and momentum at the same time. He could show this mathematically. With a lovely little equation to describe the electron's motion, Heisenberg showed that when he pinpointed momentum, position came out as indeterminate (that's math language for "none of your damn business"). When he nailed down position, momentum was indeterminate (again, math language's quiet way of giving us the finger). If you find one property, the other one blows up into nonsense.

This drove the scientific community nuts. They wanted answers. Einstein was not pleased.

"God does not play dice with the universe."

He famously protested, "God does not play dice with the universe." The electron just threw her head back, laughed, and zipped off again with an unknown momentum while the smartest men in the world suffered over their unrequited love of her. And still they suffer.

## The Physics of Life: Mystery Is Sexy

I doubt Coach Lucido knew that he was giving the class the best dating tutorial we would ever get, but here's what I got out of it: If you want attention from someone, think like an electron. Let your actions be a mystery to them. Let them know where you are but not what you are doing, or the other way around. Not both.

This is essentially what my mom had been telling me since my freshman year of high school. She already instructed me to never answer the phone all breathless and eager, but with a practiced boredom and only after several rings. When I wanted to take nonchalance to the next level, she was totally game. This is how she would answer the phone when a boy called:

BOY: Hi. Is Christy there?

MOM: Hmmm, I think so. Is this Tanner?

[NOTE: I didn't know a Tanner.]

BOY: Uh, no, this is Steve.

MOM: Oh, hi, Stan.

BOY: Steve.

MOM: Let me see if she's home yet. She had a magazine shoot after her helicopter-licensing exam.

I would then laugh in the background as though several people were at my house for a party while Mom held the phone away and said, "Oh, you're here, honey. There's a Stewart on the phone. Is he the drummer you are going to audition this weekend?" No one in high school expects moms to be liars, especially not good liars, so this worked beautifully. I learned that mystery feeds romance and fuels attention.

The reverse works nicely, too. If someone is chasing me that I'm not interested in, I tell him how I played right field on my Anchorage softball team, show him all my scars, and tell the story behind each one—every skateboarding wipeout and kitchen burn. By the time I start looking for my baby pictures, his interest has been thoroughly doused.

Thank you, electron, for this valuable lesson: Overdisclosure kills interest. Mystery fuels it.

## ((( PHYSICS PRACTICE )))

1. Ernest Rutherford determined the structure of the atom by firing alpha particles (which are equivalent to a helium nucleus) at thin gold foil and observing their deflection. A "plum pudding" distribution of matter in an atom would have allowed alpha particles to zip through the gold foil in an even distribution. Most of the alpha particles did just that, but some of the alpha particles were deflected or bounced straight back. Pretend you are Rutherford being interviewed by a reporter and answer these questions. (Fake beard optional.)

   A. What's the charge of an alpha particle?

   B. What happened to most of the alpha particles in your experiment? Why?

C. Why did some of the alpha particles deflect or bounce straight back?

D. What surprised you about the results?

Answers:

A. Since it's a helium nucleus with no electrons and helium has two protons in its nucleus, the charge of an alpha particle is 2+.

B. Most of the alpha particles shot right through the gold foil because, as it turns out, gold atoms are made of mostly nothing.

C. The positively charged alpha particles that got close to a gold nucleus (also positively charged) were thrown right back in my direction or deflected at an angle.

D. It surprised me that atoms are so roomy everywhere except the nucleus and that the nucleus is so tiny and dense.

2. If someone you like is texting you, but they have been nonspecific about their interest and intentions, how do you answer the texts below? Think like an electron and come up with appropriately mysterious answers.

A. Whatcha doin'?

B. What are you wearing?

C. Hi.

**D.** :)

**E. When will you be home?**

**F. Where are you?**

**Sample answers:**

A. Trying to remember the time difference between San Francisco and Milan.

B. Gas mask. Gotta go. Alarms.

C. Hi. Who is this?

D. Do not, in any case ever, respond to an emoticon. If someone is interested in you, they must use their words like a big boy or girl.

E. Whenever they release me on my own recognizance.

F. There are so many wonderful answers for this one. Here are some of my favorites:

   i. Backstage.

   ii. Just made it over the border.

   iii. You know you don't have the security clearance for me to release that information. (Insert frowny face or winking emoticon.)

# 20

# RESPECT OTHER POINTS OF VIEW
## RELATIVITY

Coach Lucido was talking about Newton again. Isaac felt like an old friend to us by now, a comfortable home base. We could easily picture Newton's laws because we saw them in action every day. Newton based his work on very sensible assumptions about space and time that were easy to follow. First, you can choose one point in the universe as a reference point and measure everything from there. It doesn't matter which point, exactly. It can be the North Pole of our planet, the center of the Sun, or the falafel place on Ethiopia Street in Jerusalem. Second, time keeps ticking forward no matter what else happens.

These still sound like reasonable starting points to most of us. Space and time are the predictable backgrounds upon which we change our speed and our position. We age at a consistent pace and change our watches when we land in a different time zone with the certainty that space and time do not shrink, expand, slow down, or speed up.

With space and time firmly in place, Newton was able to describe how masses collide, are accelerated by force, and stay at rest or in motion. It's all very linear and sensible. We operate as though this is true most of the time because in our daily lives, it is.

When Einstein started considering light and the speed at which it moves, Newton's vision of the universe got wobbly. Einstein's musings started out innocently enough. As a teenager, he thought about the same things we all do in our youth: When will I get my first kiss? What will I be when I grow up? What would it be like to ride a bicycle alongside a light beam? The last question took him longer to answer than the others. By the time Einstein was riding his bicycle beside imaginary light beams, it was already established that the speed of light didn't ever change. It was what we call a universal constant. Light always travels at a stubborn 299,792,458 meters per second (or 186,000 miles per second). So if young Einstein rode his bike at the speed of light and a friend turned a flashlight on as Einstein passed him, would Albert and the light move ahead into the darkness at the same speed? Would everything in front of him and the light beam remain completely dark as he kept pace with the light streaming forward?

Einstein worked through this scenario in his special theory of relativity. At this point in his career, he went from thinking about riding a bicycle next to a beam of light to thinking of traveling on a train away from or toward a beam of light. (If you read at any length about relativity, be prepared for a lot of talk about trains.) We are used to experiencing relative speeds like this: If you are in a train going really fast and you pass a herd of horses as they are running in the opposite direction, those horses are going to fly past your window faster than they would if your train wasn't moving. If the horses are running in the same direction as your train, you won't pass them as quickly. You'll have time to admire their flowing manes from your window as you pass them. From your frame of reference on the train, the speed difference between you and the horses depends on the direction in which they are running, how fast they are going, and how fast you are moving on the train. So can we expect the same thing from light? If we are traveling toward a light source, will the light beam whiz past faster than if we are not moving? If we are traveling away from the source, will light seem to slow down? No. Light will always appear to be traveling at $3 \times 10^8$ m/s.* If you are in a train going really fast away from a beam of light, it approaches you at $3 \times 10^8$ m/s. If your train is speeding toward the beam of light, it approaches you at $3 \times 10^8$ m/s.

So if we insist that light moves at the same speed no matter what, something has to give. The light beam hitting the train at the same speed every time doesn't make mathematical sense. Einstein twirled his pencil in his unruly hair and asked himself how the speed of light could remain constant from any point of view, moving or fixed. Speed is distance divided by time. Distance can't be stretched or shrunk. Time can't be stretched or shrunk. Or *can* they be? What if time doesn't tick along at the same pace in every frame of reference? What if a person flying by you near the speed of light can look at their watch and see that it

---

* Let's round up and call the speed of light 300,000,000 m/s even though light can't break the 299,792,458 speed limit.

ticks out the seconds as consistently as always, but as you watch them pass, you see the second hand on their watch is ticking off more slowly than yours? It's so crazy it just might be true, thought Einstein.

That's right; to a stationary observer, *time* slows down for you as your speed increases on that train, or on a rocket ship, or on the back of Albert's bicycle. Newton's good ol' dependable time adjusts in a way that maintains the speed of light as a universal constant. If you are clinging onto Einstein's jacket as he pedals furiously and gets as close as he can to the speed of light, your watch will slow down according to an observer standing still and watching you pass.

> To a stationary observer, *time* slows down for you as your speed increases.

Poor Newton gets all turned around when it comes to gravity, too. Instead of two bodies pulling on each other, Einstein proposed a different model. Massive bodies, like our planet, curve space like a bowling ball on a mattress. That dip in space causes smaller objects to fall toward it. This is not Newton's democratic relationship of masses. This is space-fabric bullying. Time is not fixed, and space can be warped and bent. Where will the madness end?

Your point of view is not the only correct one. While this is surprising news to any of us, it was definitely a shock to high school seniors. In Einstein's new world, all frames of reference were different and equally valid. It's one thing to say that we need to respect other people's points of view. I'd heard that in my Life Choices for Seniors class with Sister Mary, but to hear it in this way from Einstein made a bigger impression on me. It depends on where you are, how fast you are going, what mass you are near. Everything depends on everything. Truly.

## ((( PHYSICS PRACTICE )))

**1. How fast does light approach you if you are on a train going 50 miles per hour toward the light source?**

ANSWER: 186,000 miles per second.

**2. How fast does light approach you if you are on a train going 50 miles per hour away from the light source?**

ANSWER: 186,000 miles per second.

**3. How fast does light approach you if you are dancing on a Mardi Gras float that is going 3 miles per hour?**

ANSWER: 186,000 miles per second, and you put your shirt on right this very minute!

## ((( TRY THIS! )))

Because it was impossible to ride his bike anywhere near the speed of light, Einstein famously did "thought experiments." Let's do one of those:

If your twin sister (yes, you have one) took off in a snazzy single-seat rocket, did one lap around the Moon, and returned to Earth in just a few minutes, would she then be older or younger than you?

ANSWER: Because your sister had to go very fast in her rocket, relative to you the time she experienced was shorter than the time you experienced in her absence. Her watch will be slightly behind your watch and she will now tease you about being so much *older* than her.

# 21

# ENJOY THE MILES TO GO
## THE FOUR FUNDAMENTAL FORCES

If everything depends on everything, how does it all fit together? Coach Lucido was at the front of the class with chalk in hand. I was all ready to write down the equations, pencil poised. Then, in the silence that followed, I realized Coach Lucido was asking *us* that question. We were dipping into the slippery unknown and Coach was inviting us, as the next generation of thinkers, to figure out how everything we learned so far fit together.

This is exactly what Miss Randall did to us in American literature. Instead of giving us answers, she encouraged us to find the layered meanings in our reading assignments. She let us realize on our own that Willy Loman wasn't simply a pathetic salesman and Robert Frost's snowy woods weren't just a quiet place to rest. As in Coach's class, there was a lot more to it than that, and he was now challenging us to find the answers.

## The Four Forces

What's involved in this challenge? We know that there are four fundamental forces in the universe: gravitational, electromagnetic, and the strong and weak nuclear forces. Those first two forces are felt through long distances. The more distance we put between the Earth and ourselves, the less influence the Earth's gravitational pull has on us, but it still pulls. The same is true for electromagnetic push or pull. The last two forces, the strong and weak nuclear forces, are different. They are only influential in very close quarters—the nucleus of the atom.

We tend to think of gravity as the biggest deal among the forces. Early in our lives as we learn to walk, ride a bike, and couples skate, we develop a great respect for gravity. We concentrate on not falling down. We develop a healthy fear of falling for our own survival. Respect for gravity is deeply programmed in us.

We don't think much about electromagnetic forces unless we are playing with a kitchen magnet and paper clips. Even then, if you were asked which force is stronger, you would probably answer "gravity" very loudly not only because you are sure of the answer, but also so that gravity would not get annoyed and decide to throw you down the stairs later. But despite the fearsome role gravity plays in our lives, electromagnetic and the nuclear forces are much stronger. Even the "weak" nuclear force is stronger than gravity. Remember the atom that Mr. G described with all that space between the nucleus and the electrons? The force holding that atom together, despite the speedy energy of the electrons, is magnetic. The negative electrons are born to be wild and love to stay in motion, but they are magnetically attracted to the positive protons. So they stick around. Because they stick around, we have atoms. Without the magnetic force of attraction, we wouldn't have atoms or chemistry class . . . or a universe.

Since that's almost too weird to think about, here's another way to wrap your head around the strength of magnetic forces versus gravity. The next time you are traveling and you take a flying leap onto the hotel bed to check its firmness, take a moment to ponder why you didn't fall right through the bed.* You took a running start from the door, sailed through the air, and your atoms (made of mostly nothing) hit the bed's atoms (also made of mostly nothing). So why the big collision and satisfying bounce? Why didn't you sail through to the floor below you? Why didn't you fall all the way through the ground, for that matter?

The magnetic force of the atoms pushing back on each other is responsible for the impact. Your body's atoms, even as spacious as they are, can't spoon right up to the bed's atoms. The bed's atoms have protons that don't want to get anywhere near your body's protons.

In this case, electromagnetic force is stronger than gravity. Atoms are mostly space, but your atoms push back so hard on other atoms with opposing magnetic force that you can't put your hand through a wall or fall through concrete. But how do those positive protons all hang out in the

> In this case, electromagnetic force is stronger than gravity.

---

* Everyone does this, right?

nucleus together? Carbon has 6 protons all cozied up together in the nucleus. If the magnetic force is so strong, why aren't they trying to get away from each other? This is where the strong atomic force does its work. It glues the nucleus together with feisty determination. The weak nuclear force, as its name implies, is not as beefy as the strong nuclear force, but still very important. It is responsible for beta decay in which an electron (or its cousin the positron) is kicked out of an atom's nucleus and a proton converts to a neutron or vice versa.

Of the four fundamental forces, the strong nuclear force is the big daddy. Next is electromagnetic force, then the weak nuclear force, and in dead last is gravity. The challenge is to find a theory that doesn't just help us understand and predict how one works by itself, but how they all work together—a Theory of Everything.

Einstein worked on a unifying theory until his death, trying to unite and reconcile what he knew about electromagnetic and gravitational force. Physicists are working on a unifying theory still. String theory is the front-runner, with its unimaginably tiny loops and strings vibrating through multiple dimensions.

We want to believe that such a theory exists, that at some point, maybe the very first second of the universe's birth, all the forces coexisted equally and we can know how they are still entwined. It's odd that we are so confident. What makes us so sure that the universe and its forces can make sense to us? How do we know that we have the brainpower for this puzzle? We expect sensible answers from the universe, but how do we know when we've hit the limits of our understanding and we are like dogs staring at the lawn mower, wondering how it works? How do we know, too, when we have our arms around Everything with a capital E? A Theory of Everything we consider complete today might be quaint and outdated in a hundred years.

We soldier on. We can't help it. We are encouraged by all our small victories: satellites, indoor plumbing, and custard-filled doughnuts. We don't yet have a unified theory, but we know bits of the truth.

There are plenty of people who claim to have it all figured out "except for the math." That's like saying you have a sure Oscar-winning movie on your hands, you just need to write the screenplay, cast it, shoot it, and edit it. Otherwise it's totally ready. The math *is* the thing that needs to work out. It isn't simply a nagging detail. Math is the language of the universe.

We all hate it when we're in the middle of something really difficult and someone reminds us to "enjoy the journey." But those annoying people are right. The good stuff happens on the way to the goal. In trying to work out the math and falling short of the big daddy of theories we became smart enough to create transistors and computer chips. There is so much to be gained and saved on the way to perfection. We have plenty still to figure out before we understand why atoms have mass, why our universe exists at all, and how to order Thai takeout using only telepathy. On the way to figuring out the connection of the four forces we may find a way to finish Tesla's project of transmitting wireless energy.

Like Frost's traveler in the snowy woods, this is no place to stop. We have miles to go before we sleep.

# FINAL NOTE

The day I graduated from high school, Chuck stood on a chair taking pictures, and Mom smiled and shrugged to let me know there was nothing she could do to stop him. My plan was to get my engineering degree, then get a job and support my parents while Chuck went to college for his engineering degree. I imagined Chuck and I calling each other to talk about power plant construction and debating vibration suppression strategies over Thanksgiving dinner.

The first part of my plan went perfectly. I pushed my way through the mechanical engineering program in college and sent Mom and Chuck report cards so they could track my progress. Chuck called, asked questions about my classes, and helped me when I got stuck on combustion cycles or electric circuits. When I graduated, he told me, "The world is your oyster, kid."

Ten days later, Mom called at five thirty in the morning to tell me that Chuck had died. He was a ship's engineer, and while the vessel was in dock, Chuck and his crew were preparing it to head back to Alaska. There was a fire, and everyone got out but him and one other worker. Chuck was forty years old.

I fell back into a restless half sleep and dreamed that Chuck was explaining how time flexed and bowed, how he would meet me at the points in my life when I needed help. He was showing me on a blackboard with a graph, grinning like Einstein. In my dream, he'd found a way to stay with me using the laws of physics—the only laws I believed in.

Years later, in my gray high heels and wraparound dress, I felt very grown up as I walked through the courtyard of my high school. The same place Chuck had stood on his chair to take pictures of me walking single file with the other seniors in my white graduation gown. I had come back for a career fair, invited by Sister Eleanor to tell the students about engineering.

At my table, girls approached in their plaid skirts and blue sweaters, and politely asked questions for their career exploration assignments.

"What do you like about your job?"

"What advice do you have for someone entering your field?"

"What classes should I take now to prepare?"

I answered their queries with information about income, job security, and math prerequisites. I told them how good it felt to be respected for your ideas, how great it was to have the freedom to work anywhere in the world, and how much fun it was to solve real problems.

After the students returned to their classes, Sister Eleanor gave a few of us a tour of the campus. I'd heard from other people that going back to their high school made the place feel small and grubby, but it was just the opposite for me. The brick inner court, sunny lawns, and white columns glowed and hummed. I'd found my faith in that place, not the faith of mustard seeds, fig trees, and the many Marys of the New Testament, but of the gloriously precise laws of motion, energy, and gravity.

As we toured the campus, the girls sped by us holding books, talking in breathless run-on sentences. I wanted to stop them and say:

> Listen, this is important. It's hard to believe now, but your heart will be broken. Your parents will die when you still need them to take care of you. You will hide in the bathroom stall and cry at your first job when you make a stupid mistake. Nothing will go as planned. Because of life's uncertainty, you need something you can be absolutely sure of. Learn everything you can about the structure and forces of the world because you will need that solid place to stand when you are called upon to be brave, tough, and smart. Even if you don't become an engineer or a scientist, learn to think like one. Insist on what is *real*, not what you wish was real. Embrace the constraints and realities of deadlines, finances, and rejection with the same clear-eyed acceptance as you do the laws of gravity, motion, and energy. If you do that, you will become more and more capable. Then you can be fiercely and unapologetically successful in whatever you choose to do.

That's what I wanted to say, but I was listening to Sister Eleanor. She was showing us where the new gym would be built. With her arms out-

Even if you don't become an engineer or a scientist, learn to think like one.

stretched, standing on the back lawn, she announced that she wanted to swim in the new pool before she retired. "Mind, body, and spirit! We can't ignore the body!"

I was so lucky to land at my high school with the earthy and practical Sisters of St. Joseph. Even luckier to find a home in the laws of physics. Now when I see the laws play out in daily life I feel I am in on the world's secrets. Nature doesn't feel threatening, she feels like one of us—fierce, beautiful, and smart. The world is a big, overwhelming place, but when you distill it to the basic laws, it is manageable.

> The world is a big, overwhelming place, but when you distill it to the basic laws, it is manageable.

Don't try to fight physics. You are exceptional, but you are not an exception to the laws of the universe. They are bigger than you. You are planted in this time and place, with the dials of gravity, motion, and energy fixed in certain positions. You can't mess with the dials, so why try? Why not memorize their settings and work with them? Why not apply what you know about the physical world to your personal life?

Maybe you thought you needed better luck, a prettier face, or different parents. But now you know better. Everything you need is with you every day. Atoms, gravity, energy, momentum, and even quirky entropy will never surprise you by changing the rules overnight. They dance the same steps every second of your life. You know the steps now. Start dancing.

# ACKNOWLEDGMENTS

Thank you to Laurie Fox, my wonderful agent who found the perfect home for this book with Marian Lizzi at Perigee. Marian, I'm so grateful that you understood my story and were willing to include it all: gravity, entropy, Newton, Einstein, Mom, Chuck, and Sister Eleanor.

Daniel H. Wilson, thank you for introducing me to Laurie and sharing her with me.

Linda Chester, it is a pleasure to be on your team. You are truly a rock star.

Thanks to Mark Nerys, for the wonderful illustrations and for sticking with this project from proposal to final draft.

I relied on many friends for notes and feedback. They were generous with their time and gently let me know when I sounded like a weirdo. Among those friends were Courtenay Hameister, Susan Galluccio, Laurel Long, Julie Livingston, Jon Manning, Beth

Morris, John Breen, Tanya Wuertz, Sadie Bryan, and John Rydzewski. An extra special thanks to Courtenay Hameister and Robyn Tenenbaum for inviting me to be on *Live Wire*. You and your audience showed me that there were other people who thought the laws of thermodynamics were touching and hilarious.

Thanks to the smart folks I work with in the engineering and construction world who have allowed me to work on their team while giving me the freedom to write, make music, and shoot television: Karl Shulz, Bart Eberwein, Keith Krueger, Creighton Kearns, Matt Jackson, and Paul Giorgio.

Big swan hugs to my creative soul mates Lara Michell and Stephanie Schneiderman, who started this physics conversation with me years ago during a long drive on tour, then wisely sent me to the backseat when the lack of sleep got to me and I thought the gas station was serving piping hot revenge.

Thank you to Scott Reis for welcoming me into his math classes to tutor his brilliant students at De La Salle in North Portland. The experience reminded me of the importance of great teachers like you.

I am so grateful to all the wonderful teachers who showed me such kindness and worked hard to keep me busy at College Gate Elementary and Wendler Junior High in Anchorage and at Carondelet High School in Concord.

Thank you to my sister, Cathy, and my mom, Kathryn, for always listening to my ideas and laughing at my jokes. A big kiss to Greg for his endless encouragement and detailed knowledge of long-distance sniper shooting.

Thank you most of all to Chuck McKinley, who wholeheartedly took on the job of being my dad. All I ever want is to make you proud.

# INDEX

Page numbers in *italics* indicate figures; those followed by "n" indicate notes.

# ABOUT THE AUTHOR

*Photo by Alicia Rose*

**Christine McKinley** grew up in Anchorage, Alaska, and earned her BS in mechanical engineering at California Polytechnic University in San Luis Obispo. She is a licensed mechanical engineer in Oregon and California. She has hosted television shows on the Discovery and History channels. She lives in Portland, Oregon, where she plays bass and guitar in the band Swan Sovereign. Her musical about physics, *Gracie and the Atom*, won a Portland Drammy for Original Score. Visit her website at christinemckinley.com.